ANNALS *of* THE NEW YORK ACADEMY OF SCIENCES

VOLUME
1276

ISSUE

Annals Meeting Reports

TABLE OF CONTENTS

1 Dissecting signaling and functions of adhesion G protein–coupled receptors
Demet Araç, Gabriela Aust, Davide Calebiro, Felix B. Engel, Caroline Formstone, André Goffinet, Jörg Hamann, Robert J. Kittel, Ines Liebscher, Hsi-Hsien Lin, Kelly R. Monk, Alexander Petrenko, Xianhua Piao, Simone Prömel, Helgi B. Schiöth, Thue W. Schwartz, Martin Stacey, Yuri A. Ushkaryov, Manja Wobus, Uwe Wolfrum, Lei Xu, and Tobias Langenhan

26 Scientific considerations for complex drugs in light of established and emerging regulatory guidance
Chris Holloway, Jan Mueller-Berghaus, Beatriz Silva Lima, Sau (Larry) Lee, Janet S. Wyatt, J. Michael Nicholas, and Daan J.A. Crommelin

37 Fetal programming and environmental exposures: implications for prenatal care and preterm birth
Thaddeus T. Schug, Adrian Erlebacher, Sarah Leibowitz, Liang Ma, Louis J. Muglia, Oliver J. Rando, John M. Rogers, Roberto Romero, Frederick S. vom Saal, and David L. Wise

ANNALS *of* THE NEW YORK ACADEMY OF SCIENCES

EDITOR-IN-CHIEF
Douglas Braaten

ASSOCIATE EDITOR
Rebecca E. Cooney

PROJECT MANAGER
Steven E. Bohall

Artwork and design by Ash Ayman Shairzay

The New York Academy of Sciences
7 World Trade Center
250 Greenwich Street, 40th Floor
New York, NY 10007-2157

annals@nyas.org
www.nyas.org/annals

**The New York
Academy of Sciences**

Published by Blackwell Publishing
On behalf of the New York Academy of Sciences

Boston, Massachusetts
2012

Ann. N.Y. Acad. Sci. ISSN 0077-8923

ANNALS OF THE NEW YORK ACADEMY OF SCIENCES

Issue: Annals *Meeting Reports*

Dissecting signaling and functions of adhesion G protein–coupled receptors

Demet Araç,[1] Gabriela Aust,[2] Davide Calebiro,[3] Felix B. Engel,[4] Caroline Formstone,[5] André Goffinet,[6] Jörg Hamann,[7] Robert J. Kittel,[8] Ines Liebscher,[9] Hsi-Hsien Lin,[10] Kelly R. Monk,[11] Alexander Petrenko,[12] Xianhua Piao,[13] Simone Prömel,[9] Helgi B. Schiöth,[14] Thue W. Schwartz,[15] Martin Stacey,[16] Yuri A. Ushkaryov,[17] Manja Wobus,[18] Uwe Wolfrum,[19] Lei Xu,[20] and Tobias Langenhan[8]

[1]Stanford University, Stanford, California. [2]Department of Surgery, Research Laboratories, University of Leipzig, Leipzig, Germany. [3]Institute of Pharmacology and Rudolf Virchow Center, DFG-Research Center for Experimental Biomedicine, University of Würzburg, Würzburg, Germany. [4]Department of Cardiac Development and Remodelling, Max-Planck-Institute for Heart and Lung Research, Bad Nauheim, Germany, and Laboratory of Experimental Renal and Cardiovascular Research, Department of Nephropathology, Institute of Pathology, University of Erlangen-Nürnberg, Erlangen, Germany. [5]MRC Centre for Developmental Neurobiology, King's College London, New Hunts House, London, United Kingdom. [6]Université Catholique de Louvain, Institute of Neuroscience, Developmental Neurobiology, Brussels, Belgium. [7]Department of Experimental Immunology, Academic Medical Center, University of Amsterdam, Amsterdam, The Netherlands. [8]Institute of Physiology, Department of Neurophysiology, University of Würzburg, Würzburg, Germany. [9]Institute of Biochemistry, Molecular Biochemistry, Medical Faculty, University of Leipzig, Leipzig, Germany. [10]Department of Microbiology and Immunology, College of Medicine, Chang Gung University, Tao-Yuan, Taiwan. [11]Department of Developmental Biology, Washington University School of Medicine, St. Louis, Missouri. [12]Shemyakin-Ovchinnikov Institute of Bioorganic Chemistry, Moscow, Russia. [13]Division of Newborn Medicine, Department of Medicine, Boston Children's Hospital and Harvard Medical School, Boston, Massachusetts. [14]Department of Neuroscience, Functional Pharmacology, Uppsala University, Uppsala, Sweden. [15]Laboratory for Molecular Pharmacology, Department of Neuroscience and Pharmacology, and the Novo Nordisk Foundation Center for Basic Metabolic Research, University of Copenhagen, Copenhagen, Denmark. [16]Faculty of Biological Sciences, University of Leeds, Leeds, United Kingdom. [17]Division of Cell and Molecular Biology, Imperial College London, London, United Kingdom, and Medway School of Pharmacy, University of Kent, Chatham, United Kingdom. [18]Medical Clinic and Policlinic I, University Hospital Carl Gustav Carus, Dresden, Germany. [19]Cell and Matrix Biology, Institute of Zoology, Johannes Gutenberg University of Mainz, Mainz, Germany. [20]University of Rochester Medical Center, Rochester, New York

Address for correspondence: Tobias Langenhan, Institute of Physiology, Department of Neurophysiology, University of Würzburg, Röntgenring 9, 97070 Würzburg, Germany. tobias.langenhan@uni-wuerzburg.de

G protein–coupled receptors (GPCRs) comprise an expanded superfamily of receptors in the human genome. Adhesion class G protein–coupled receptors (adhesion-GPCRs) form the second largest class of GPCRs. Despite the abundance, size, molecular structure, and functions in facilitating cell and matrix contacts in a variety of organ systems, adhesion-GPCRs are by far the most poorly understood GPCR class. Adhesion-GPCRs possess a unique molecular structure, with extended N-termini containing various adhesion domains. In addition, many adhesion-GPCRs are autoproteolytically cleaved into an N-terminal fragment (NTF, NT, α-subunit) and C-terminal fragment (CTF, CT, β-subunit) at a conserved GPCR autoproteolysis–inducing (GAIN) domain that contains a GPCR proteolysis site (GPS). These two features distinguish adhesion-GPCRs from other GPCR classes. Though active research on adhesion-GPCRs in diverse areas, such as immunity, neuroscience, and development and tumor biology has been intensified in the recent years, the general biological and pharmacological properties of adhesion-GPCRs are not well known, and they have not yet been used for biomedical purposes. The "6th International Adhesion-GPCR Workshop," held at the Institute of Physiology of the University of Würzburg on September 6–8, 2012, assembled a majority of the investigators currently actively pursuing research on adhesion-GPCRs, including scientists from laboratories in Europe, the United States, and Asia. The meeting featured the nascent mechanistic understanding of the molecular events driving the signal transduction of adhesion-GPCRs, novel models to evaluate their functions, and evidence for their involvement in human disease.

Keywords: G protein–coupled receptors; GPS motif; autoproteolysis; molecular and genetic analysis

doi: 10.1111/j.1749-6632.2012.06820.x

Introduction

The biennial adhesion-GPCR workshops evolved from a European grassroots initiative that began in 2002 and have been fostering informal communication on topics concerning adhesion-GPCR research. The workshops slowly evolved from gatherings of a group of immunologists working on the founding adhesion-GPCR class members, F4/80 and CD97, to the only international meeting solely dedicated to adhesion-GPCR research. This year the tenth anniversary workshop gave more than 20 laboratories a platform on which to exchange their latest insights into the biology of this poorly understood receptor class. At the same occasion, the Adhesion-GPCR Consortium (AGC) was formed by the constituting member assembly (http://www.adhesiongpcr.org). The AGC will represent and disseminate adhesion-GPCR research and coordinate concerted funding initiatives on the topic. The workshop program was divided into (1) evolutionary aspects of adhesion-GPCRs, (2) signaling of adhesion-GPCRs, (3) adhesion-GPCRs in development, (4) adhesion-GPCRs in neurobiology, and (5) adhesion-GPCRs in disease.

The origin of the adhesion-GPCR family

Helgi Schiöth (Uppsala University) introduced his studies on the origin of the adhesion-GPCR class of seven-span transmembrane (7TM) receptors. The adhesion G protein–coupled receptors (GPCRs) are the second largest family of GPCRs, with genes containing multiple exons that are presumed to have arisen through several mechanisms.[1,2] Adhesion-GPCRs are of ancient origin and are found in several eukaryotes that include most of the vertebrates, the closest relatives to the vertebrates (*Ciona intestinalis* and *Branchiostoma floridae*), and the most primitive animals (*Nematostella vectensis* and *Trichoplax adhaerens*; Fig. 1).[3] Intriguingly, gene mining in amphioxus *B. floridae* has revealed several novel adhesion-GPCR domains such as somatomedin B, kringle, lectin C-type, SRCR, LDLa, immunoglobulin I-set, CUB, and TNFR, typically not found in the mammalian receptors.[4] Further, unique adhesion-GPCRs have been identified in urochordates (*C. intestinalis*) and in the *Strongylocentrotus purpuratus* (sea urchin) genome, and some of these are species specific. There are at least 21 adhesion-GPCRs in *C. intestinalis* that possess just the GPCR proteolysis site (GPS) proteolytic domain in the N-termini, while in the sea urchin there are 40 adhesion-GPCRs containing multiple leucine-rich repeats (LRRs) but lacking GPS sites.[5,6] Furthermore, comprehensive analysis of the entire set of adhesion and related secretin, and Methuselah groups of GPCRs provided the first evolutionary hierarchy among the five main classes of vertebrate GPCRs. Schiöth's group provided convincing evidence that the secretin GPCRs descended from the family of adhesion-GPCRs, probably from group V of the adhesion-GPCRs.[7] Moreover, they clarified the origin of the adhesion-GPCRs by providing the first evidence for the presence of adhesion-GPCR homologues in fungi.[8] This study estimated that the adhesion-GPCRs evolved from *Dictyostelium* cAMP receptors before the split of unikonts from a common ancestor of all major eukaryotic lineages.[8] In addition, they mined the close unicellular relatives of the metazoan lineage *Salpingoeca rosetta* and *Capsaspora owczarzaki*. These species have a rich group of the adhesion-GPCRs that provided additional insight into the first emergence of the N-terminal domains of the adhesion family.[8] Prime examples are the emergence of the characteristic adhesion-family domains, GPS and the Calx-β domain in *C. owczarzaki*, and the EGF-CA domain in *S. rosetta*.[8] Further, Schiöth analyzed the hemichordate *Saccoglossus kowalevskii* (acorn worm), which serves as an important model organism for developmental biologists to understand the evolution of the central nervous system (CNS). Unlike vertebrates that have a centralized nervous system, the acorn worm has a diffuse nerve net. Despite this, the acorn worm contains well-conserved orthologues for several of the adhesion family members with a similar N-terminal domain architecture. This is particularly apparent for those genes responsible for CNS development and regulation in vertebrates (Krishnan, A., *et al.* unpublished data). Overall, adhesion-GPCRs have a remarkably long and complex evolutionary history that can be traced down to the common ancestors of metazoa and fungi (Fig. 1). Knowledge of the origin and evolution of the unique N-terminal domain architectures of these genes may offer opportunities to better understand their functional roles and to aid deorphanization.

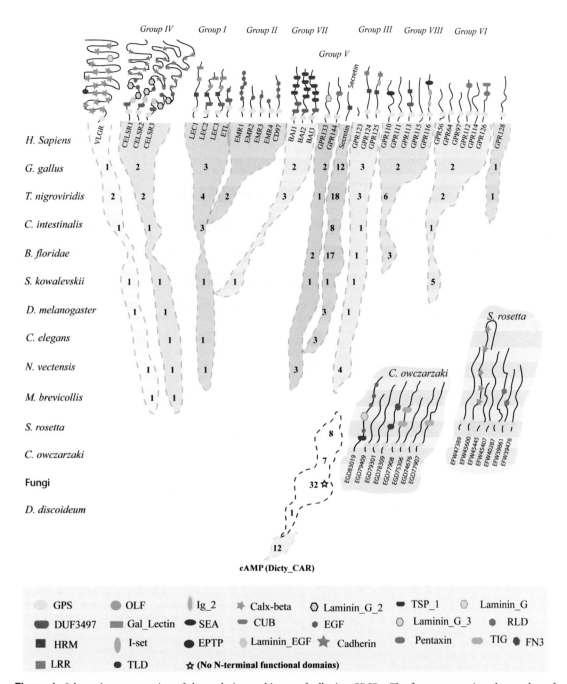

Figure 1. Schematic representation of the evolutionary history of adhesion-GPCRs. The figure summarizes the number of adhesion-GPCR family sequences found in each group across species. Each row denotes a species, and the colored fields are the adhesion-GPCR groups. N-terminal domains are shown for the human adhesion-GPCR genes. The N-terminal domain architecture of the unclassified novel adhesion-GPCRs found in *C. owczarzaki* and *S. rosetta* is shown separately. The red-colored star represents the absence of N-terminal domains for the novel fungi homologues.

Signaling of adhesion-GPCRs

Intense focus during the 6th Adhesion-GPCR Workshop was directed toward the mechanistic understanding of adhesion-GPCR signal transduction in relation to the structural components of adhesion-GPCRs. A group of researchers presented their findings on details of the molecular signaling mechanism of different adhesion-GPCRs, providing a comprehensive overview of methods, models, and receptors currently used to understand adhesion-GPCR signal transduction.

The GPS motif: 15 years of studies

Alexander Petrenko (the Shemyakin-Ovchinnikov Institute of the Russian Academy of Sciences), summarized previous research into the structural hallmark of adhesion-GPCRs and the GPS, which lies at the heart of connecting receptor structure with function. Purification and molecular cloning of the calcium-independent receptor of latrotoxin (CIRL/latrophilin/CL) revealed the unusual composition of two heterologous subunits that derive from the endogenous cleavage of the precursor protein at the extracellularly oriented site close to the first transmembrane segment of the CIRL heptahelical core. A cysteine-box motif surrounding this site of cleavage appeared to be conserved in the adhesion-GPCR family and in a few adhesion-like single-span transmembrane proteins. Similarly to the CIRL, all analyzed proteins with this motif appeared to be proteolyzed. Moreover, when mutations were introduced in this region, receptors were no longer cleaved. They therefore named this motif the *GPCR proteolysis site*, or GPS.[9] In addition to four cysteine residues, the GPS motif also contains two conserved tryptophans and strong preferences in other residues within the motif. GPS-defined cleavage of the CIRL and other adhesion-GPCRs takes place in the endoplasmic reticulum, thus avoiding furin-mediated proteolysis. Two major issues were addressed in further studies of the GPS in the laboratories of Petrenko and others. The first issue was the mechanism of the proteolysis; the second was the structure of the resulting protein products. For the CIRL, Petrenko and colleagues showed that the presence of the GPS motif was necessary for cleavage. With truncated mutants, they demonstrated the cleavage of a large CIRL fragment containing the GPS and a neighboring latrotoxin-binding domain, which is weakly homologous to the corresponding

region in adhesion-GPCR BAI. However, they failed to observe cleavage of the recombinant protein with the GPS motif only. No protease involved in the GPS cleavage has been identified. On the basis of the presence of the *cis*-proteolysis signature, it was proposed that the cleavage is autocatalytic.[10] Yet, in an *in vitro* system they found that GPS cleavage can be regulated indicating a role for either a protease or some chemical cofactor. In their original CIRL description, they showed that the two cleavage products, p120 and p85, are tightly bound. However, later, p120 and p85 were found to overlap but only partially, thus suggesting their independent localization on the membrane.[11] They observed a separate soluble p120 fragment, but it was due to a secondary extracellular cleavage of the complex. Interestingly, the two-subunit complex can be dissociated under harsh conditions *in vitro*, but the fragments will not reassociate. Also, an extracellular peptide fragment of p120 that binds to p120 in native CIRL complexes would not bind to recombinant p120. All these observations received explanations in a recently published study on the structure of cleaved and uncleaved GPS-containing protein fragments of CIRL and BAI.[12] The GPS motif is part of a larger fold that was named the *GAIN domain*. Within this domain, the fragment with the GPS motif is tightly embedded into a larger structure, and there are no major structural changes or dissociation upon the cleavage. Future studies will have to address further the functional significance of GPS proteolysis and clarify its mechanism.

A novel and evolutionarily conserved domain of adhesion-GPCRs mediates autoproteolysis

Demet Araç (Stanford University) introduced her work on the structural elucidation of the GAIN domain. Unlike other GPCRs, adhesion-GPCRs have large extracellular regions that are autoproteolytically cleaved from their seven-pass transmembrane regions at a conserved GPS.[9] Previously, it was believed that the so-called stalk region that precedes the GPS motif of all adhesion-GPCRs was nonfunctional and unstructured. Unexpectedly, Araç discovered that the GPS motif itself does not constitute an autonomously folded domain, but rather forms a single folded domain together with the so-called stalk region. Thus, the ~40 residue GPS motif is an integral part of a much larger ~320 residue domain that they termed the *GPCR-Autoproteolysis INducing*

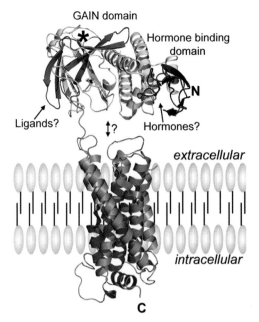

Figure 2. All adhesion-GPCRs have an extracellular GAIN domain that precedes the first transmembrane helix. The GAIN subdomain A is colored yellow. The GAIN subdomain B is colored light pink. The GPS motif, which is part of subdomain B, is colored magenta. The HormR domain (which exists in 12 of human adhesion-GPCRs) is colored blue. The modeled transmembrane helices are colored orange. The cleavage site is indicated with an asterisk. Possible interactions are indicated with a question mark.

(GAIN) *domain.* Araç and colleagues determined the crystal structures of GAIN domains from two distantly related adhesion-GPCRs, CL1 and BAI3, and revealed a conserved novel fold that was previously unidentified (Fig. 2).[12]

Strikingly, the GAIN domain is the only extracellular domain shared by all 33 human adhesion-GPCRs and all five human polycystic kidney disease proteins. Database searches have revealed that primitive organisms, such as *Dictyostelium discoideum* that arose early in evolution before animals emerged, encode GAIN domains, although they lack most other autoproteolytic domains, important adhesion and signaling domains, and critical signaling pathways. These findings show that the GAIN domain is a widespread and conserved autoproteolytic domain in higher eukaryotes as well as in ancient organisms.

Functionally, the entire GAIN domain is both necessary and sufficient for autoproteolysis, as determined by deletion experiments of CL1. Araç *et al.*

performed extensive mutagenesis of the CL1 cleavage site and revealed the unique structural features of the GAIN domain that enable self-cleavage. Autoproteolysis occurs between the last two β-strands of the GAIN domain in a short and kinked loop, suggesting an autoproteolytic mechanism whereby the overall GAIN domain fine-tunes the chemical environment in the GPS to catalyze peptide bond hydrolysis. The GAIN domain is the locus of multiple human disease mutations, including cancer, autosomal dominant polycystic kidney disease, and bilateral frontoparietal polymicrogyria. The disease-causing mutations on the GAIN domains may interfere with autoproteolysis function or other functions of the GAIN domain such as ligand binding.

Two properties of the GAIN domain–mediated autoproteolysis make it unique and intriguing. First, all GAIN domains always immediately precede the N-terminal transmembrane helix by a short linker and are in close association with the signaling transmembrane domains (Fig. 2). Second, in contrast to most other autoproteolytic domains, upon autoproteolysis, the GAIN domain remains attached to the membrane-embedded regions of the protein. These observations naturally lead to the hypothesis that the GAIN domain may regulate receptor signaling via the transmembrane helices; and GAIN domain–mediated autoproteolysis has a complex mechanism of action. Indeed, deletion experiments, reported by others, indicate that the GAIN domain may have an inhibitory role on GPCR signaling.[13]

The hormone receptor (HormR) domain is the second most frequently observed domain in adhesion-GPCRs (found in 12 of 33 human adhesion-GPCRs). However, no hormones have yet been identified to bind adhesion-GPCRs. Remarkably, Araç's HormR domain crystal structures from CL1 and BAI3 revealed an unusually high structural similarity (0.7 Å r.m.s.d.) to the genuine hormone-binding domain of the corticotrophin-releasing factor receptor (CRFR), suggesting that adhesion-GPCRs may be hormone receptors.

In summary, Araç's work has redefined the poorly understood adhesion-GPCR class, showing that members of this family share a large, unique, and widespread autoproteolytic domain that may be involved in downstream signaling and in human disease.

Activation of adhesion-GPCRs by an endogenous tethered ligand

Simone Prömel (University of Leipzig) introduced her studies on the GPS motif of latrophilins, which have shed light on the molecular interactions that the GAIN/GPS element of adhesion-GPCRs participates in. Latrophilins (LPHN/CL/CIRL) have been characterized to be one of the receptors for α-latrotoxin, a component of the black widow spider toxin.[14] Binding of α-latrotoxin to LPHN1 leads to calcium-independent release of neurotransmitters in neurons.[15] Besides CELSR/Stan/Flamingo, latrophilins are the only members of the adhesion-GPCR family that are present in vertebrates and invertebrates. They can be found throughout several phyla and species, making them prototypic for the adhesion-GPCR class and suggesting that they have essential roles that are conserved in all bilaterians. Three latrophilin homologs exist in the mammalian genome (*lphn1–3*), whereas the nematode *Caenorhabditis elegans* contains two homologs, *lat-1* and *lat-2*. A *lat-1* null mutant causes variable morphogenic defects during embryonic and larval development, leading to embryonic and larval lethality. Previous studies indicate that LAT-1 participates in the control tissue polarity during embryogenesis.[16]

The molecular mechanisms of LAT-1 signaling remain elusive. Very few *in vivo* approaches to address such questions have been developed thus far. However, Prömel and colleagues employed an *in vivo* assay in *C. elegans* to assess molecular function of the latrophilin homolog LAT-1. The assay is based on transgenic rescue of *lat-1* mutant phenotypes by a wild-type or a modified *lat-1* copy, which allows analysis of structure–function relationships in the biological context of the receptor without detailed knowledge of input or output signals.

Using this assay, Prömel performed a comprehensive receptor analysis indicating that LAT-1 signals via two different types of interactions.[17] The first mode requires the presence and structural preservation of the GPS—which is an integral part of the GAIN domain[12]—and 7TM domains. By contrast, the other interaction is independent from the 7TM/C terminus of the receptor. Importantly, for both modes of receptor activity the GPS is essential. On a more mechanistic level, Prömel and colleagues uncovered that the GAIN/GPS domain mediates receptor activity by interacting with the 7TM

of the receptor, consistent with its function as an endogenous tethered ligand of the 7TM domain (Fig. 3). This finding suggests that the GAIN/GPS structure might exert a similar function among several, if not all, adhesion-GPCRs by modulating the signaling of the 7TM domain.[17] Further, the assay system allowed intermolecular complementation experiments with pairs of LAT-1 receptor variants *in vivo*. These experiments revealed that the GPS cross-interacts with the 7TM domain of a homologous partner receptor, likely in a dimeric complex, for the 7TM-dependent function (Fig. 3). Prömel also tested the requirement of GPS proteolysis, which cleaves latrophilins and most other adhesion-GPCRs autoproteolytically.[10] Thus far this cleavage event has been assumed to be essential for receptor function.[18] In contrast to that assumption, Prömel and colleagues showed that GPS cleavage is not essential for receptor function.[17] Finally, the GAIN/GPS structure is required for a second function in *C. elegans* fertility that operates independently of the 7TM domain, indicating the molecular versatility of the GAIN/GPS region in adhesion-GPCRs (Fig. 3).[17]

The work by Prömel describes novel insights into adhesion-GPCR function on a molecular level and analyses under *in vivo* conditions. These insights, based on a complete receptor analysis, are the first steps toward a better understanding of the entire class of adhesion-GPCRs and a general mechanism for their mode of signaling. In future studies both modes of receptor activity, 7TM-dependent and -independent, and their impact on latrophilin function, will be analyzed in more detail. Additionally, current research by Prömel addresses the question of how GPS and the seven-transmembrane domain might interact to mediate function and which other interaction partners play a role.

G protein–mediated signal transduction of adhesion-GPCRs

Ines Liebscher (University of Leipzig) demonstrated a high-throughput approach to determine G protein–mediated signal transduction of adhesion-GPCRs. Over the last several years attempts have been made to unravel the issue of signal transduction of adhesion-GPCRs. Although several ligands as interacting partners with these receptors had been identified, activation of specific intracellular signal cascades remained obscure. There were

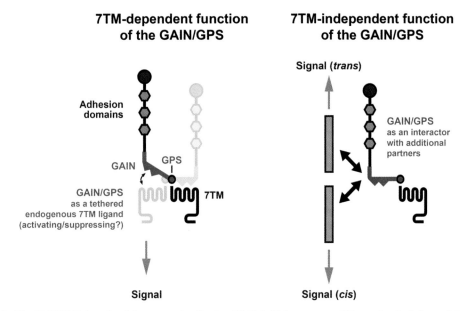

Figure 3. The GAIN/GPS domain of the nematode adhesion-GPCR LAT-1 serves two different signals. Left panel: in the 7TM-dependent mode, the GAIN/GPS domain functions as a tethered endogenous ligand of the 7TM domain. This interaction could occur in a dimeric complex of two homonymous receptor molecules, where 7TM the domain is cross-activated by the GAIN/GPS of the partner molecule, a mechanism akin to the activation of receptor tyrosine kinases. Right panel: in the 7TM-independent mode, the GAIN/GPS interacts with additional partners to transduce a separate signal (forward or reverse).

only a few reports on intracellular signaling mechanisms of adhesion-GPCRs. It was shown that latrophilin 1, the prototypic adhesion-GPCR, induced intracellular Ca^{2+} signaling upon interaction with the exogenous ligand α-latrotoxin.[19] GPR56 appeared to activate the $G\alpha_{12/13}$ protein/Rho-pathway after stimulation with an antibody against the ectodomain.[20] BAI1 recognized phosphatidylserine and could directly recruit a Rac-GEF complex to mediate the uptake of apoptotic cells.[21] However, clear evidence of intracellular signaling for most adhesion-GPCR via G proteins is still missing.

Bohnekamp and Schöneberg have recently shown that overexpression of the adhesion-GPCR GPR133, which is associated with adult height and the RR interval duration in an electrocardiogram, activates $G\alpha_s$, leading to an increase of cAMP levels.[22] $G\alpha_s$ protein–coupling of the basally active GPR133 was verified by $G\alpha_s$ knockdown with siRNA, overexpression of $G\alpha_s$, co-expression of a chimeric $G\alpha_{qs4}$ protein that routes receptor activity to the phospholipase C/inositol phosphate pathway, and by missense mutation within the transmembrane domain. Liebscher's data provided strong evidence to suggest that this member of the adhesion-GPCR family functionally interacts with the $G\alpha_s$/adenylyl signal-

ing cascade. Further analysis showed that the presence of the N terminus and the cleavage at the GPS are not required for G protein signaling of GPR133. Liebscher has extended these investigations to other family members in order to study both the generality and specificity of G protein–mediated signal transduction of adhesion-GPCRs. Preliminary data indicate that GPR116, GPR123, GPR124, and GPR126 also couple to the $G\alpha_s$/adenylyl cyclase pathway. GPR115, GPR116, and GPR126 appear to activate $G\alpha_i$, and GPR115 is the only adhesion-GPCR so far that couples to the $G\alpha_q$/phospholipase C pathway. Recently, $G\alpha_s$ protein coupling was verified for GPR114 and GPR133, whereas GPR97 showed $G\alpha_o$ coupling.[23] These new data prove that classical receptor/G protein interaction is a common feature of adhesion-GPCR signaling.

A tethered inverse agonist model for activation of adhesion-GPCRs expressed on enteroendocrine cells

Thue Schwartz (University of Copenhagen and the Novo Nordisk Foundation Center for Basic Metabolic Research) described the impact of adhesion-GPCR signaling on enteroendocrine cells, and added to the theme of a potential molecular signaling mechanism of adhesion-GPCRs. Enteroendocrine cells function as specialized sensors of food components and nutrient metabolites. The cells are flask shaped with dense core secretory granules located at the base, from which peptide hormones are released, and with an apical microvillus–decorated sensory extension reaching the gut lumen.[24] Much attention has recently focused on the expression and function of 7TM receptors as chemosensors for metabolites, for example, of triglycerides—long chain fatty acids and 2-OG—and of complex carbohydrates generated by the gut microbiota—short chain fatty acids.[24,25] The enteroendocrine cells are, along with their neighboring enterocytes, renewed every week from pluripotent stem cells located at the bottom of the mucosal crypts.

Individual enteroendocrine cells are isolated and FACS purified after genetic labeling with GFP or RFP expressed under the control of promoters for gut hormones such as CCK.[26] Through quantitative polymerase chain reaction (qPCR) analysis, classical nutrient receptors were identified as being both highly expressed and highly enriched in the cells. Surprisingly, a number of adhesion-GPCRs were identified as being expressed at similar high levels in enteroendocrine cells as were specific nutrient metabolite receptors, for example, Celrs1, GPR128, Celsr3, CD97, and Lphn1. However, the majority of the adhesion-GPCRs were also highly expressed in the neighboring enterocytes, though a few were found to be both highly expressed and highly enriched in the enteroendocrine cells.

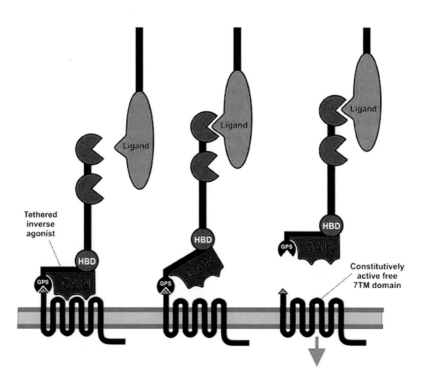

Figure 4. General model for molecular activation mechanism for adhesion-GPCRs. On the basis of molecular pharmacological studies on adhesion-GPCRs expressed on enteroendocrine cells, it is proposed that the large N-terminal extension in the full-length version of these receptors acts as a tethered inverse agonist, which, by binding to the 7TM domain, inhibits or silences the otherwise high constitutive activity of this domain. Upon binding of one or more of the far N-terminal ligand-binding domains (indicated in purple) to their macromolecular ligand, located either in *trans* on an opposing cell, in *cis* on the same cell, or in the intercellular matrix, the tethered inverse agonist (indicated in red) is removed from the 7TM domain, either partly (middle panel) or totally, which is possible in adhesion-GPCRs, provided that the GAIN/GPS domain–mediated autocleavage has occurred (right panel). Through this ligand binding–mediated process the free 7TM domain of the adhesion-GPCR will start signaling with its high constitutive activity, which, in this way, in fact, functions as a ligand/agonist-mediated signaling.

                                    Ann. N.Y. Acad. Sci. 1276 (2012) 1–25 © 2012 New York Academy of Sciences.

It has been rather unclear to what degree adhesion-GPCRs couple through classical G protein pathways. A number of the adhesion-GPCRs expressed in enteroendocrine cells were cloned and expressed heterologously in HEK293 cells using an optimized signal peptide construct to improve expression levels, though generally they displayed only minimal signaling. However, strong signaling through both G_q and G_i and, in particular, the SRE transcriptional activation pathway—that is, presumably through $G_{12/13}$—was observed when the receptors were expressed in an N-terminally truncated form in which only the small N-terminal segment from the autocleavage site to TM-I was intact. Thus, a general activation model for adhesion-GPCRs was proposed (see Fig. 4). In which the 7TM domain of the adhesion-GPCRs is highly constitutively active and the large N-terminal segment of the receptors functions as a tethered inverse agonist. That is, in the intact receptor, the N-terminal extension, or presumably the 3D-conserved GAIN domain, which after the autocleavage is noncovalently bound to the 7TM domain, will silence the constitutive signaling of this domain. It is proposed that binding of one or more of the far N-terminally located binding domains (which differ among the different adhesion-GPCRs) to a ligand attached on the neighboring cell or located on the same cell or in the intercellular matrix will lead to dissociation of the N-terminally tethered inverse agonist, which results in high constitutive signaling of the unbound 7TM domain left at the cell surface (Fig. 4).

This model is in agreement with a similar model suggested by the Hall and Xu groups on the basis of classical biochemical structure–function studies of GPR56, and by a model proposed by Langenhan and coworkers on the basis of *in vivo* structure–function studies performed in *C. elegans* with LAT-1/latrophilin.[13,17,27]

Shear stress–dependent downregulation of CD97 on circulating leukocytes by CD55

Jörg Hamann (University of Amsterdam) presented work on the ligand interactions of CD97, a prototypic adhesion-GPCR broadly expressed by hematopoietic and nonhematopoietic cells. CD97 interacts, through different regions in its extracellular subunit, with at least four other molecules: CD55, chondroitin sulfate B, $\alpha_5\beta_1$ integrin, and CD90/Thy-1. The ability of CD97 to engage with seemingly unrelated binding partners has triggered studies that aim to address the importance of individual ligands *in vivo* using the interaction with CD55 (Ref. 28) as a paradigm. These studies revealed that mice lacking either CD97 or CD55 had higher granulopoietic activity, resulting in increased numbers of circulating granulocytes.[29] Moreover, the absence of CD97 or CD55 reduced disease activity in two experimental models of arthritis.[30] In both cases, CD97 and CD55 knockout mice developed a highly similar phenotype; yet a causative relationship between the molecules could not be established.

Hamann described that circulating leukocytes from CD55-deficient mice express significantly increased levels of CD97. After adoptive transfer into of CD55-deficient leukocytes wild-type mice, CD97 expression on CD55-deficient leukocytes dropped to normal levels due to contact with CD55 expressed on wild-type leukocytes and stromal cells. Downregulation of CD97 occurred within minutes after first contact with CD55, involved both the extracellular and transmembrane subunit of the receptor, and correlated with an increase in plasma levels of soluble CD97. *In vitro*, downregulation of CD97 on CD55-deficient leukocytes cocultured with wild-type blood cells was strictly dependent on the shear stress from rigorous agitation of the cell cultures. *In vivo*, CD55-mediated downregulation of CD97 required intact circulation, as shown in experiments with wild-type recipient mice that were pretreated with heparin to prevent blood coagulation and then sacrificed immediately after adoptive transfer, followed by blood collection at later time points; transferred CD55-deficient leukocytes did not downregulate CD97 under these conditions. To test whether ligation by CD55 triggers CD97 signaling, CD55-deficient leukocytes were cocultured with wild-type blood cells. Notably, *de novo* ligation did not activate signaling molecules that recently were shown to be constitutively engaged by CD97 in cancer cells, such as ERK, PKB/Akt, and RhoA.[31]

Taken together, the findings presented confirm CD55 as a genuine binding partner of CD97 *in vivo*. They suggest that CD55 downregulates CD97 surface expression on circulating leukocytes by a process that requires physical forces but does not, based on current evidence, induce receptor signaling (Fig. 5). Regulation of CD97 expression by CD55 may prevent uncontrolled clustering of leukocytes

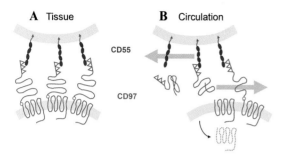

Figure 5. Consequences of the CD97–CD55 interaction *in vivo*. (A) In tissue, contacts between the adhesion-GPCR CD97 and its ligand CD55 likely facilitate cell adhesion. (B) In the circulation, CD97 expression is constantly downregulated by contact with CD55 on blood and stromal cells. This regulation process may prevent uncontrolled clustering of leukocytes in the blood stream, thereby restricting CD97–CD55 interaction-mediated adhesion to tissue sites.

due to homo- or heterotypic cellular contacts in the blood stream, thereby restricting CD97–CD55 interaction-mediated adhesion to tissue sites. The data support the hypothesis that adhesion-GPCRs are two-part entities with distinct roles for the extracellular and the seven-transmembrane subunits in cell adhesion and signaling, respectively.

Activation of the EMR2 receptor via ligation-induced translocation, and interaction of receptor subunits in lipid rafts activates macrophages

Hsi-Hsien Lin (Chang Gung University) analyzed the signaling mechanism of the adhesion-GPCR EMR2 in macrophages. Directed migration of phagocytes to infected sites is a critical step in innate immunity for pathogen elimination. Activated phagocytes clear invading pathogens by multiple mechanisms, including phagocytosis and release of proteases, antimicrobial peptides, and cytokine/chemokines. As a myeloid cell–restricted member of the adhesion-GPCR family, the EMR2 receptor has been shown previously to play a role in the cellular functions of innate immune cells.[32,33] Indeed, ligation of the EMR2 receptor not only can increase neutrophil adhesion and migration, but it can also augment the production of antimicrobial mediators.[32,33]

As with the majority of adhesion-GPCRs, EMR2 is posttranslationally modified by GPS autoproteolysis in the endoplasmic reticulum and cleaved into a large extracellular domain (α-subunit) and

a seven-transmembrane domain (β-subunit).[10,34] To investigate the role of GPS autoproteolysis in mediating the cellular functions of adhesion-GPCRs and the mechanistic relevance of the receptor subunit interaction, Lin and coworkers first demonstrated that GPS proteolysis is necessary for EMR2-mediated cell migration. Next, the structural organization of EMR2 receptor subunits was examined. Surprisingly, two distinct receptor complexes were identified: one is a noncovalent α-β heterodimer, while the other consists of two independent receptor subunits with differential distribution in lipid raft microdomains. More specifically, the EMR2 α-subunit was shown to locate mostly in the nonraft regions, while the β-subunit was found in both the raft and nonraft regions. These data suggest that the two EMR2 receptor subunits do not always interact on the cell surface but behave, in part, as two independent molecules.[35]

Moreover, they showed that ligation of the EMR2 receptor by the α-subunit–specific 2A1 monoclonal antibody induces the translocation and colocalization of receptor subunits in lipid rafts. Such ligation activated the EMR2 receptor in macrophages, leading to the production of inflammatory cytokines such as IL-8 and TNF-α. Interestingly, cytokine production was inhibited when macrophages were treated with lipid raft disruptors lovastatin and filipin.[35] Thus, EMR2 receptor ligation–induced cytokine production seems to require intact lipid raft microdomains. Recently, preliminary data from Lin and coworkers indicated that EMR2 receptor activation via ligation of receptor subunits induces the activation of extracellular signal–regulated kinase (ERK), leading to macrophage activation. The induction of ERK phosphorylation in macrophages is 2A1 specific and dose and time dependent. Furthermore, ERK phosphorylation and cytokine production by macrophages via EMR2 receptor ligation was inhibited by U0126, an MEK inhibitor. Finally, supernatant of macrophages stimulated by 2A1-mediated EMR2 receptor ligation was found to stimulate human neutrophil activation, promote its transwell migration, and augment fMLP-induced ROS production. Taken together, these results demonstrate that the EMR2 receptor plays a critical role in innate immune functions (Fig. 6), and provides a paradigm for signal transduction within the adhesion-GPCR family.

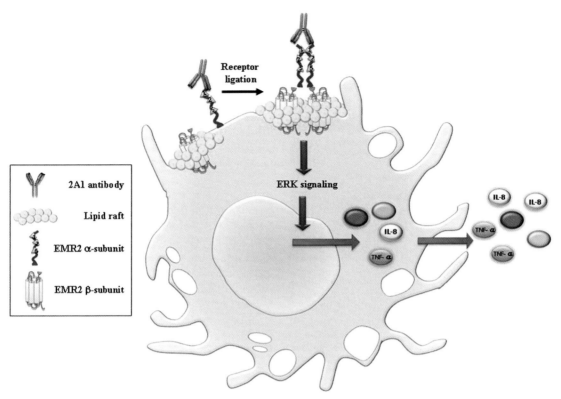

Figure 6. Activation of the EMR2 receptor in macrophages is mediated through the translocation and interaction of its two independent subunits in the lipid raft. The independent EMR2 α- and β-subunits are localized in the nonraft and lipid raft regions, respectively. Following receptor ligation by the α subunit–specific 2A1 mAb, the independent α-subunit translocates and reassociates with the β-subunit in the lipid raft region. Such interaction induces intracellular signaling via the MAPK pathway (mainly ERK1/2 phosphorylation), leading eventually to proinflammatory cytokine (IL-8, TNF-α) secretion.

Real-time monitoring of GPCR signaling in living cells

Davide Calebiro (University of Würzburg) concluded the workshop section on signaling of adhesion-GPCRs with an overview of *in vivo* imaging for GPCR signaling, which might be extended to adhesion-GPCRs one day. The approximately 1000 different GPCRs present on the surface of cells provide fundamental links between the extracellular environment and the intracellular milieu, allowing cells to respond and adapt to a wide variety of stimuli, such as hormones, neurotransmitters, light, and odors, as well as cell and matrix contacts. Whereas the basic molecular mechanisms of GPCR signaling have been elucidated, how such diverse stimuli are integrated via a few common signaling cascades while achieving highly specific responses is still poorly understood. In the last few years, Calebiro and colleagues developed a series of optical methods using fluorescence resonance energy transfer

(FRET), which enables imaging GPCR activation and signaling directly in living cells.[36]

More recently, in order to analyze GPCR signaling under highly physiological conditions, Calebiro and colleagues have generated a transgenic mouse[37] with ubiquitous expression of a FRET sensor for cAMP[38] (Fig. 7A). This mouse has allowed them to study, among other aspects, the signals produced by the activation of a prototypical hormone receptor, that is, thyroid-stimulating hormone receptor (TSHR), directly in intact thyroid follicles (Fig. 7B). Unexpectedly, the results indicate that the TSHR, and possibly other GPCRs, can continue stimulating cAMP production even after internalization into the endosomal compartment, which leads to persistent signaling (Fig. 7C) and specific effects.[37,39] These data reveal new and important functions for receptor internalization in regulating GPCR-mediated responses. Calebiro and colleagues are currently using similar approaches with FRET sensors to further

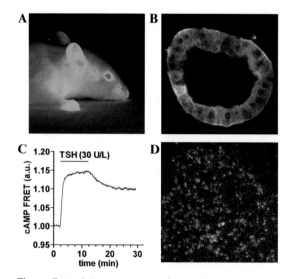

Figure 7. Real-time monitoring of GPCR signaling in living cells. (A) Transgenic mouse with ubiquitous expression of a FRET sensor for cAMP (Epac1-camps). (B) Confocal image of a thyroid follicle isolated from the cAMP reporter mouse. (C) Representative FRET trace obtained in a thyroid follicle, showing persistent cAMP elevations after transient TSH stimulation. (D) Single GPCRs on the surface of living cells visualized by TIRF microscopy.

explore this and other novel aspects of GPCR-cAMP signaling in fundamental physiological processes such as thyroid hormone production and female reproduction. Whereas these approaches allow a precise characterization of GPCR and second messenger signaling with high spatiotemporal resolution, a full characterization of GPCR signaling cascades will likely require observing the signals produced by the activation of a single receptor.

To achieve this goal, Calebiro and colleagues are developing new methods using labeling with small organic fluorophores and total internal reflection fluorescence (TIRF) microscopy, which allow visualizing signaling proteins at the surface of living cells with single-molecule sensitivity (Fig 7D). They are using these methods to monitor individual protein–protein interactions, such as those involved in ligand binding, receptor di-/oligomerization, or coupling to G proteins with high spatiotemporal resolution. Initial data suggest that GPCRs are targeted to different microdomains of the cell surface, where they are present in a dynamic equilibrium, with constant formation and dissociation of new receptor complexes. Taken together, these data provide novel insights into the complex dynamic events at the basis of the spatiotemporal compartmentalization of GPCR signaling cascades.

Adhesion-GPCRs in development

Accumulating evidence demonstrates that adhesion-GPCRs fulfill plentiful functions during the genesis of various organ systems. For some adhesion-GPCRs, these functions are well defined on a cell biological level, whereas other receptors remain to be placed in a physiological context.

Molecular and genetic analysis of Gpr126 in peripheral nerve development

Kelly Monk (Washington University School of Medicine) found a role for Gpr126 in myelin formation. Myelin is the multilayered glial membrane surrounding axons in the vertebrate nervous system. Myelin acts as an insulator that reduces the capacitance of the axonal membrane, thus increasing axon conduction velocity and allowing for fast processing speeds and efficient transmission of information over large distances. Myelin not only insulates axons: the glial cells that make myelin also protect and provide vital trophic support to neurons, and the importance of myelin is underscored in diseases in which it is disrupted, like multiple sclerosis and peripheral neuropathy. In the peripheral nervous system (PNS), Schwann cells produce myelin by iteratively wrapping their membranes around axons. Reciprocal signaling between axons and Schwann cells is required for proper myelination to occur, but the exact signaling mechanisms regulating myelination are poorly understood. An orphan adhesion-GPCR, Gpr126, is required for the initiation of myelination in zebrafish Schwann cells.[40] In the mouse, loss of *Gpr126* leads to complete amyelination in the PNS (Fig. 8) as well as multiple defects in peripheral nerves.[41] In zebrafish, Monk previously showed that forskolin treatment to elevate cAMP levels suppresses the mutant phenotype and restores myelination,[40] and they hypothesized that elevation of cAMP levels would similarly suppress mutant phenotypes in mouse mutants. Indeed, cAMP elevation or protein kinase A (PKA) activation in dorsal root ganglion explant cultures from *Gpr126* mutant mice rescued the myelin defects, providing further support that cAMP and PKA are involved in the Gpr126-mediated pathway initiating myelination. Although *Gpr126* is predominantly

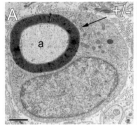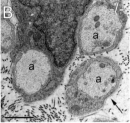

Figure 8. Gpr126 is essential for Schwann cell myelination. (A) An axon (a) is surrounded by myelin (arrow) in a postnatal day 12 *Gpr126*[+/−] sciatic nerve. (B) Schwann cells (arrow) ensheathe axons (a) in postnatal day 12 *Gpr126*[−/−] sciatic nerve, but do not make myelin. Scale bars = 1 μm.

expressed in Schwann cells, systemic *Gpr126* knockouts show pronounced axon loss. Therefore, Monk's group has begun to analyze conditional *Gpr126* mouse mutants, and their preliminary analysis suggests that Gpr126 is required autonomously in Schwann cells for myelination in mammals.

Discrepancies in Gpr126 knockout phenotypes

Felix Engel's laboratory (Max-Planck-Institute for Heart and Lung Research) performed a large-scale temporal mRNA expression analysis describing rat heart development from embryonic day (E) 11 to postnatal day 10 in an interval of 12 hours. This study identified the adhesion-GPCR *gpr126* as a gene that is transiently expressed during embryonic rat heart development. These data suggested that Gpr126 might be important for heart development. Moreover, it has recently been demonstrated that *gpr126* knockout mice exhibit a thinned myocardial wall. In addition, *gpr126* deletion has been described as embryonically lethal between E10.5 and E12.5 days post fertilization.[42] Preliminary data in zebrafish (in Engel's laboratory) further substantiated this hypothesis. By contrast, Monk *et al.* did not describe a heart phenotype or embryonic lethality in *gpr126* mutant zebrafish (*gpr126*[st49]). Instead, they describe an ear phenotype and found that Gpr126 is essential for Schwann cells to initiate myelination.[40] What could explain the discrepancies between the GPR126 knockout mouse and knockdown or mutant zebrafish phenotypes?

Little is known about Gpr126. As for most adhesion-GPCRs there is no identified natural or synthetic ligand of Gpr126. Gpr126 contains a seven-transmembrane domain (7TM), here called

the C-terminal fragment (CTF), and an extracellular domain containing a CUB (complement, Uegf, Bmp1) domain and a pentraxin (PTX) domain,[43] here called the N-terminal fragment (NTF). The NTF and the CTF are linked via a GPS-containing stalk region, which is highly conserved among adhesion-GPCRs. Adhesion-GPCRs appear to be cleaved at the GPS immediately after biosynthesis, but NTF and CTF are generally believed to stay noncovalently associated after cleavage and to be expressed on the cell membrane as a heterodimer. Importantly, Volynski *et al.* showed that the NTF of the adhesion-GPCR latrophilin can be self-anchored to the cell membrane independently of the CTF.[11] Currently, it is unclear whether NTFs of other adhesion-GPCRs can exist independently and what physiological importance they have. However, Moriguchi *et al.* have previously demonstrated that Gpr126 (DREG) is also cleaved and that a small fragment is secreted.[44] Interestingly, the mutation in the *gpr126*[st49] fish introduces a stop codon just before the GPS domain.[40] Thus, it is likely that *gpr126*[st49] mutant fish might express a functional NTF. Therefore, Engel hypothesizes that the NTF of Gpr126 functions independently of its CTF and that the NTF, but not the CTF, is required for the developing heart. To test this hypothesis Engel and colleagues are performing selective knockdowns of CTF or the entire *gpr126* in zebrafish. Moreover, they will analyze whether overexpression of NTF in Gpr126 morphants can rescue the heart but not the ear phenotype. Finally, they are in process to generate conditional knockout mice to better determine cell type–specific functions of Gpr126.

Role of Celsr1–3 cadherins in planar cell polarity and ependymal development

André Goffinet (Université Catholique de Louvain) researches the CELSR group of adhesion-GPCRs. Cadherin EGF LAG seven-pass G-type receptors 1, 2, and 3 (Celsr1–3) form a family of three atypical cadherins with multiple functions in epithelia and the nervous system. During the last few years evidence has accumulated for important and distinct roles of Celsr1–3, as well as other genes, such as Frizzled3 (Fzd3), in planar cell polarity (PCP) and brain development and maintenance.[45,46] Celsr1–3 harbor large ectodomains, composed of nine N-terminal cadherin repeats, EGF-like domains, laminin G repeats, one hormone receptor motif, and a potential GPS. This is followed by seven trans-membrane domains and a cytoplasmic tail. Like their fruit fly ortholog Flamingo (*Fmi*), Celsr1–3 are thought to work in interaction with other core PCP proteins such as Frizzled (particularly Fzd3 and Fzd6 in mammals), Van Gogh (Vangl1–2), Dishevelled (Dvl1–3), and Prickle (Pk1–4).

Observations of constitutive and conditional Celsr2 and 3 mutant mice uncovered important functions of these proteins during ependymal development. In double mutant animals, a severe hydrocephalus develops rapidly during the early postnatal period (P), leading to death around P8–P10. Studies with scanning and transmission microscopy and immunohistochemistry found prominent abnormalities of ciliogenesis in mutants. Basal bodies remained frequently embedded in the subapical cytoplasm and failed to become aligned normally (rotational polarity defect). As a result, many ependymal cells are completely devoid of cilia and the circulation of the CSF is severely impaired, leading to lethal hydrocephalus.[46]

More recently, Goffinet has studied the role of Celsr1 during ependymal development using conditional and floxed Celsr1[47] and Vangl2[48] mutant alleles, showing that they are distinct from that of Celsr2 and 3. Whereas cilia appear at the normal time and are of normal length, in Celsr1 mutant mice there is a clear defect in translational polarity, in that cilia tufts are not consistently displaced rostrally, as in normal animals; a translational polarity defect was seen even at P0, when ependymal cells had one nonmotile cilium, in Celsr1 mutants and in Fzd3 and Vangl2 mutants. Rotational polarization of basal bodies (BB) was studied using double labeling with gamma-catenin for BB and phosphocatenin, which labels an area adjacent to BB, opposite the basal foot. Rotational polarization of BB was found to be similarly defective in Celsr1 and Vangl2 mutants (Fzd3 mutants die at P0 and cannot be studied). In addition, the mean vectors of translational and rotational polarization are not aligned in mutant ependymal cells, unlike normal ependymal cells.

Goffinet's data show that Celsr1 regulates translational and rotational polarity of ependymal cells from the time that they are generated to the adult stage. Celsr1 works together with Vangl2 and Fzd3 in this process. Such PCP regulation in the ependymal epithelium is complementary to the role of Celsr2

and 3, which regulate ependymal differentiation and ciliogenesis.

Adhesion-GPCR Celsr1 in the complex morphogenesis of mammalian organ primordia

Caroline Formstone (King's College London) reported about additional functions of Celsr-like adhesion-GPCRs. Mice with disrupted core PCP component function die at birth owing to catastrophic developmental defects in neural tube closure[49] and lung branching.[50] Defects in the development of other organ systems are also apparent including the epidermis.[51] The adhesion-GPCR Flamingo plays a central role in the local transmission of PCP information among neighboring cells. Of three Flamingo homologues in mammals, Celsr1 is predominantly found in epithelial precursors within organ primordia. Several studies strongly indicate a major role for Celsr1 in the coordination of PCP during mammalian organ development,[47,51] but how it functions to coordinate epithelial morphogenesis is unclear. Formstone's recent data suggest that Celsr1 protein exhibits a differential distribution along the apicobasal axis of some epithelia. In particular, Celsr1 exhibits a novel enrichment to the basal membrane of neuroepithelial precursors and lung tubules.[52,53] Studies on how differential Celsr1 protein distribution links to its function in tissue morphogenesis and whether Celsr1 protein at the epithelial basal membrane elicits PCP signaling will provide insight into the complex roles of Celsr1 in mammalian organ development.

CD97 overexpression induces a megaintestine

Interestingly, CD97 is also implicated in intestinal development, as presented by Gabriela Aust (Universtiy of Leipzig). Adhesion-GPCRs are involved in adhesion, guidance, and positioning of cells. CD97, in contrast to the other EGF-TM7 adhesion-GPCR subfamily members restricted to immune cells, is present in normal and malignant epithelial cells. In normal human intestine, CD97 is located in enterocytic cell–cell contacts[54] and, in the cytoplasm, shows an expression gradient along the crypt–villus axis.

To understand the role of CD97 in intestinal physiology, Aust *et al.* generated transgenic Tg(villin-CD97) mice.[54] Unexpectedly, overexpression of CD97 resulted in upper megaintestine, depending on the CD97 cDNA copy number integrated. Intestinal enlargement involved an increase in length, diameter, and weight. Remarkably, the megaintestine phenotype develops with normal microscopic morphology, and thereby clearly differs from existing megaintestine models in which intestinal enlargement is often accompanied by dramatic morphological changes.

The megaintestine phenotype is acquired after birth before weaning, which makes this a unique model for investigating the mechanisms underlying postnatal expansion of the mammalian small intestine by way of two consecutive growth patterns: (1) cylindrical growth in length and diameter without alteration of microscopic morphology (as seen in the Tg(villin-CD97) mice), and (2) luminal growth with amplification of the internal surface area.[55]

Notably and in accordance with a cylindrical growth pattern, suckling but not adult Tg(villin-CD97) mice showed more crypt fission compared with wild-type mice. Consistently, acquisition of the megaintestine was independent of altered cell lineage determination, Wnt signaling, and an increase of intestinal stem cell markers. Suckling Tg(villin-CD97) pups developed the phenotype independent of the genotype of the feeding dam, thus excluding regulation of a milk growth factor by CD97. Most likely, CD97 regulates the binding or signaling of an intestinal receptor for a milk constituent.

The Tg(villin-CD97) mice provide new evidence supporting the conclusion that adhesion-GPCRs have distinct functions that may depend on the cellular context in which a given receptor is expressed. Tg(villin-CD97) mice not expressing CD55 that binds to the extracellular EGF-like domains of CD97 also developed a megaintestine, suggesting that the adhesive extracellular part is not necessary for phenotype induction. By contrast, mice overexpressing a truncated CD97 with only the first two transmembrane helices did not develop a megaintestine, which implies signaling through CD97 in phenotype induction.

Overall, these transgenic mice provide suitable models to uncover and understand functions of adhesion GPCRs in normal epithelial cells.

Neurobiological roles of adhesion-GPCRs

The connection between adhesion-GPCRs and neural function was postulated early on. Some groups reported about their ongoing efforts to pinpoint the exact relationship between these receptors and the properties of this highly specialized cell type.

Drosophila *synapses as an* in vivo *model to study structure–function relationships of latrophilin*

Tobias Langenhan and Robert Kittel (University of Würzburg) introduced the fruit fly *Drosophila melanogaster* as a new *in vivo* model for research on evolutionarily conserved adhesion-GPCRs of the latrophilin group, which is extendable to the Flamingo/CELSR group. Latrophilins have been implicated in the control of synaptic transmission as well as planar cell polarity, raising the questions of whether and how these phenomena are interlinked.[15,16,56] Thus far, models in which cell polarity and neuronal exocytosis could be tested at the same time and in the same cell type have been lacking in the adhesion-GPCR field. The versatile model system of *Drosophila* includes high-throughput transgenesis with single copy integration, homologous recombination for knockout/knockin studies of selected target genes, cell- and time-specific transgene expression through binary expression systems, and a vast arsenal of allelic variants covering the entire genome for genetic interaction studies. In particular, the fruit fly larva possesses a well-defined synaptic contact, the neuromuscular junction (NMJ), ideally suited for investigating adhesion-GPCR expression and function with biochemical, imaging, and electrophysiological methods. Langenhan and Kittel presented preliminary data indicating that latrophilin/dCIRL is resident at the NMJ, and that its removal by mutation or RNA interference causes changes in the molecular, structural, and functional properties of this synapse type. After full characterization of the phenotypic profile due to *dCirl* deficiency, Langenhan and Kittel will use the NMJ as an *in vivo* platform to test modified *dCirl* variants and correlate molecular lesions in the receptor with functional consequences on a cell biological level.

Transsynaptic interaction between presynaptic latrophilin and postsynaptic Lasso

Yuri Ushkaryov (University of Kent) is interested in the function of mammalian latrophilins and presented his latest data. Latrophilin 1 (LPH1),[57] a neuronal adhesion G-protein–coupled receptor that binds α-latrotoxin, is implicated in control of presynaptic Ca^{2+} and in the modulation of neurotransmitter release.[58,59] To understand the molecular mechanisms of these physiological functions, Ushkaryov's group isolated the endogenous ligand of LPH1, Lasso.[60] This protein is a splice variant of teneurin-2. Teneurins are brain-specific, orphan, cell surface receptors with functions in neuronal pathfinding and synaptogenesis. Ushkaryov's data indicate that LPH1, located on presynaptic terminals, forms strong and specific transsynaptic complexes with Lasso, which is found on postsynaptic spines. This interaction is not only structural but also functional: soluble fragments of Lasso induce intracellular Ca^{2+} signals upon binding to LPH1 in presynaptic boutons of cultured hippocampal neurons and in nonneuronal cells expressing exogenous LPH1. Furthermore, the LPH1–Lasso complexes play an important role in synaptic development and activity. Thus, Lasso fragments acting via LPH1 strongly increase the rate of spontaneous exocytosis in mouse neuromuscular junctions.[61] LPH1 expressed on nonneuronal cells induces postsynaptic differentiation in cocultured hippocampal neurons.[60] On the other hand, while synapses in which the interaction between LPH1 and Lasso is inhibited, appear morphologically normal, they remain physiologically silent.[61] Taken together, the data from the Ushkaryov group indicate that while the transsynaptic interaction of LPH1 and Lasso is not necessary for the initial establishment of central synapses, it participates in presynaptic Ca^{2+} control and is required for functional maturation of presynaptic nerve terminals.

The very large G PCR Vlgr1b/GPR98 – a key component of the Usher syndrome protein networks

Uwe Wolfrum (University of Mainz) described the role of the very large G protein–coupled receptor-1 (VLGR1) in the inner ear and in retinal biology. VLGR1, also known as MASS1 or GPR98, has a molecular weight of up to ∼700 kDa and is by far the largest GPCR and the largest cell surface protein known to date.[62] The large ectodomain of the largest splice variant VLGR1b contains several repeated motifs, including calcium binding, Calx-β repeats, and seven copies of an epitempin repeat. It is linked to the 7TM moiety via a proteolytic site (GPS) containing a region typical for adhesion-GPCRs. The short intracellular C-terminus contains a consensus PDZ binding motif, suggesting interac-

tions with cellular scaffold proteins. In the absence of any known ligand VLGR1/GPR98 is one of the few adhesion-GPCRs in which mutations are disease relevant. VLGR1/GPR98 defects are thought to be associated with epilepsy. Mouse *vlgr1* mutants are characterized by the susceptibility to audiogenic seizures and to the development of sensoneuronal defects, namely hearing impairment and visual dysfunction.[63,64] Mutations in the human VLGR1/GPR98 gene cause Usher syndrome (USH) type 2C.[65]

Human USH is the most common form of combined hereditary deaf-blindness. Three clinical subtypes (USH1–3) are differentiated on the basis of severity, age of onset, and progression of the symptom.[65] Wolfrum and others have identified VLGR1/GPR98 as a component of USH protein networks in inner ear hair cells and retinal photoreceptor cells. In hair cells VLGR1/GPR98 is part of the ankle link complex essential for the formation of the ankle links between the membranes of neighboring stereocilia and thereby for the correct development of the mechanosensitive hair

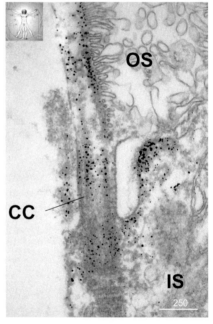

Figure 9. Immunoelectronmicroscopy localization of VLGR1/GPR98 in a photoreceptor cell of the human retina. VLGR1/GPR98 is located in the connecting cilium (CC) and periciliary region of the rod cell inner segment (IS), as well as along the axoneme of the outer segment (OS). Bar = 250 nm.

bundles.[63] In photoreceptor cells VLGR1/GPR98 is a component of the periciliary USH protein network, which is crucial for cargo transport to the photoreceptor cilium (Fig. 9).[64] In this periciliary network, VLGR1/GPR98 is required for the assembly of fibrous links communicating between the membranes of the inner segment and the connecting cilium of photoreceptor cells. In both sensory systems, VLGR1/GPR98 is additionally found at synapses, where it is specifically localized in postsynapses of the dendritic tips of retinal bipolar cells and in spiral ganglion neuritis, respectively.[66]

The identification of further components of these protein networks, the decipherment of the downstream cellular signaling pathway, and knowledge about ligands of VLGR1/GPR98 will lead not only to a better understanding of protein function, but will also enlighten the pathomechanisms underlying the USH disease, which is a necessary prerequisite for the development of future therapy concepts.

GPR56-dependent development of the frontal cerebral cortex

Xianhua Piao (Harvard Medical School) investigated the function of the adhesion-GPCR GPR56 in neural development. Although the human cerebral cortex is subdivided into dozens of specific areas with divergent functions, the genetic mechanisms underlying the regional development of the cerebral cortex are very poorly understood. One approach to studying the mechanisms of cortical specification is genetic analysis of inherited conditions in which specific regions of the cortex are preferentially perturbed. Bilateral frontoparietal polymicrogyria (BFPP), a recessively inherited genetic disorder of human cerebral cortical development, shows severely abnormal architecture in the frontal lobes, with milder involvement of the posterior parts of the cortex (Fig. 10A and B).[67] Linkage analysis and positional cloning in a cohort of 22 BFPP patients revealed that GPR56 is the causative gene of BFPP.[68] This discovery demonstrated a novel signaling pathway in the developmental regulation of regionalization of the cerebral cortex.

Unraveling the ligand of GPR56 is the first step in revealing the signaling pathway of GPR56. A receptor affinity probe *in situ* approach demonstrated that the putative ligand of GPR56 is expressed in the meninges and pial basement membrane (BM). Subsequent proteomic and genetic studies identi-

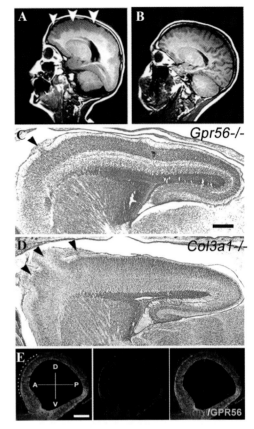

Figure 10. The gradient expression of GPR56 in the preplate correlates with the anatomical distribution of cortical defects associated with mutations in GPR56 and Col3a1. Compared to the MRI of a normal brain (B), multiple small gyri with a scalloped appearance of the cortical–white matter junction were predominantly seen in the frontal cortex in a BFPP brain (white arrowheads in A). This regional deformation of the cerebral cortex (black arrowheads) was recapitulated in Gpr56 (C) as well as in Col3a1 (D) knockout mouse brains. Interestingly, the restricted expression of GPR56 (green) at the basal surface of embryonic day 10.5 mouse brain (outlined by white dotted line in E, left panel) matches the anatomical distribution of the cortical defects seen in both humans and mice when GPR56 or its ligand collagen III is deleted. A–P and D–V axes are shown in E. Abbreviations: A, anterior; P, posterior; D, dorsal; V, ventral. This figure is adapted from previous publications as follows: A and B from Ref. 68; C from Ref. 69; E from Ref. 72.

fied collagen III as the endogenous ligand of GPR56 in the developing brain.[69] Upon binding to collagen III, GPR56 activates RhoA via coupling to $G\alpha12/13$. RhoA activation has been shown to regulate cell migration. To study GPR56-mediated RhoA activation on neuronal cell migration, an *in vitro* neurosphere migration assay was performed. The presence of collagen III inhibits neuronal migration in a GPR56-dependent fashion. This observation was further confirmed in *Gpr56* and *Col3a1* knockout mouse brains (Fig. 10C and D).[70],[71] Taken together, Piao's data indicate that the interaction of GPR56 and its ligand collagen III inhibits migrating neurons from breaking through the pial BM, thus conveying a positional cue during cortical development.

Because the regulation of rostral cortical development by GPR56 signaling could be accomplished by regional expression of either GPR56 or its ligand collagen III, Piao and coworkers studied the expression profile of both proteins in the developing cortex. Immunohistochemistry of collagen III on sagittal sections of mouse embryonic brains ranging in age from E10.5 to E11.5 did not reveal an expression gradient of collagen III during these developmental stages.[72] In contrast, an anterior-to-posterior gradient of GPR56 protein expression was found on the basal surface of the neocortex in both E10.5 and E11.5 brains (Fig. 10E), but dissipated by E12.5. This finding is particularly interesting, as the change in the expression pattern occurs in the region where preplate neurons reside.

During cerebral cortical development first-born neurons form the preplate directly beneath the pial BM and function as a framework for further development of the cortex. However, the molecular mechanism underlying the function of the preplate neurons remains largely unknown. The fact that a gradient expression of GPR56 in preplate neurons matches the regional cortical defects associated with loss of GPR56, or its ligand collagen III, (Fig. 10) suggests that a novel receptor–ligand pair is responsible for mediating the interaction between preplate neurons and the pial BM, thus defining the boundary between the neocortex and the meninges, while providing a framework for the developing cortex. Further testing of this hypothesis will undoubtedly advance our understanding of the molecular mechanisms underlying how preplate neurons regulate cortical development.

Emerging roles of adhesion-GPCRs in disease

It becomes increasingly clear that adhesion-GPCR dysfunction is involved in several human conditions. Research into these pathological states not only helps elucidate molecular breaking points of adhesion-GPCR signals, but also assists in developing remedies and direct pharmacological efforts to counteract adhesion-GPCR–dependent diseases. At the workshop, the roles of adhesion-GPCRs in tumorigenesis were discussed.

GPR56 and cancer

Lei Xu (University of Rochester Medical Center) presented studies from her lab on the roles of the adhesion-GPCR GPR56 in cancer progression. She and colleagues previously showed that GPR56 is downregulated in highly metastatic melanoma cells and that its re-expression led to inhibition of metastasis and melanoma growth.[73] Insights on how this might occur came from the identification of a putative ligand of GPR56, TG2.[73] TG2, also called tissue transglutaminase, is a crosslinking enzyme in the extracellular matrix (ECM) that modulates ECM biophysical properties.[74] TG2 also possesses crosslinking-independent functions and interacts with integrins and ECM proteins to regulate cell adhesion.[75] The signaling mechanisms of GPR56, as well as whether and how TG2–GPR56 interaction regulates melanoma progression, are outstanding questions for which progress was then discussed.

Initial lines of investigation revealed that angiogenesis is impaired in tumors expressing high levels of GPR56.[27] Angiogenesis, the process of nascent blood vessel formation, is essential for sustained tumor growth. Overexpression of GPR56 in the melanoma cell line MC-1 inhibited production of vascular endothelial growth factor (VEGF), a major contributor to angiogenesis, and resulted in a decrease in angiogenesis and tumor growth. In contrast, deletion of a $\sim$70 aa serine threonine proline–rich (STP) segment in the α-subunit of GPR56 resulted in a significant elevation of VEGF production and enhanced angiogenesis and melanoma growth. The opposite effects of GPR56 and ΔSTP-GPR56 in melanoma cells indicate that the seven transmembrane domains of GPR56 (GPR56β) might exist in different activation states, and that these states might be modulated by the α-subunit and/or its binding partners (Fig. 11). Consistent with this model, addition of purified GPR56α was sufficient to inhibit VEGF production from MC-1 cells expressing GPR56β.[27]

Since the STP segment in GPR56 was both necessary and sufficient for binding to TG2, the opposing effects of ΔSTP-GPR56 and full-length GPR56 on

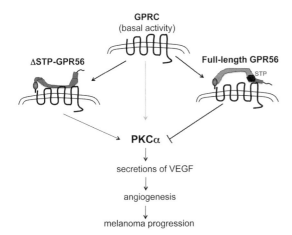

Figure 11. Different activation states of GPR56.

VEGF production implied that the GPR56–TG2 interaction was required for regulation of VEGF production by GPR56. Nevertheless, in contrast to the effect of ΔSTP-GPR56, knockdown of TG2 by shRNAs did not result in elevated VEGF production, indicating that the roles of the TG2–GPR56 interaction in melanomas might be more complex. To investigate this further, researchers analyzed growth of melanoma cells expressing GPR56 cDNA or shRNAs in immunodeficient *Tg2*[−/−] mice. Preliminary data suggested an unexpected antagonistic relationship between GPR56 and TG2 in melanomas: while GPR56 inhibited melanoma growth, TG2 promoted it (work in progress). Furthermore, the absence of TG2 abolished the effects of GPR56 on melanoma growth, indicating that TG2 might act downstream of GPR56, i.e., GPR56 might inhibit the tumor-promoting role of TG2. The mechanisms of this antagonism were revealed through a series of immunohistochemical and biochemical analyses. GPR56 expression was found to induce changes in the distribution pattern of TG2 in melanomas, probably due to a downregulation of TG2 in the ECM of GPR56-expressing melanoma cells. This downregulation was confirmed by *in vitro* studies that showed that the extracellular TG2 was internalized by GPR56 and subsequently degraded intracellularly in a lysosome-dependent manner.

Aberrant adhesion-GPCR expression in breast cancer—a potential role in metastasis?

Martin Stacey (University of Leeds) reported on the connection between adhesion-GPCRs and metastasis development. EGF-TM7 adhesion-GPCRs are predominantly expressed on leukocytes, including macrophages, dendritic cells, and neutrophils. Through the use of a stimulating antibody (2A1), Stacey and colleagues have shown that ligation of the human-restricted EGF-TM7 receptor EMR2 results in the enhanced activation of human neutrophils. Data demonstrate an increase in reactive oxygen species generation, degranulation (myeloperoxidase), and surface marker expression (CD11b and L-selectin shedding) upon ligation by 2A1. Furthermore, EMR2 transfectants displayed an increase in *in vitro* cell migration and invasion.[33] Truncations of the transmembrane domains and mutants preventing cleavage at the GPS site demonstrated the requirement of an intact transmembrane domain and receptor processing to elicit cell signaling,[33] showing that signaling is indeed required for EMR2 function. Overall, the data suggest an important role in the activation and migration of human leukocytes. Interestingly despite its leukocyte-restricted profile of EMR2, Stacey and colleagues show that mRNA and protein are aberrantly present in epithelial cells of breast cancer tissue.[76] Moreover receptor isoform expression is similar to that seen in neutrophils and macrophages, suggesting a potential hijacking of the normal function of EMR2 for tumor activation, migration, and progression. qPCR and flow cytometry analysis of EMR2-transfected breast cancer cell lines demonstrated increased expression of the epithelial-to-mesenchymal transcription factors snail and twist and decreased expression of the epithelial marker E-cadherin. Further, potential roles of EMR2 in tumor progression are to be investigated. Tools for targeting of EGF-TM7 receptors have been generated through the use of recombinant antibody fragments; for example, scFv and diabodies of antibodies against EMR2 and F4/80 have been cloned and fused to either toxins or model peptides. These reagents will be used in future depletion studies and in the receptor-specific targeting of antigens to leukocyte subsets.

The potential role of CD97 in the biology of acute myeloid leukemia

Manja Wobus (University Hospital, Dresden) found that the adhesion-GPCR CD97 is involved in acute myeloid leukemia (AML). AML cells home to a specified region of the bone marrow (BM), where they interact with stromal components, including extracellular matrix proteins, glycosaminoglycans, and stromal cells, by which they derive proliferative and growth inhibitory signals. Different receptors, for example, VLA- (very late antigen-) 4, CXCR4, and CD44, described to play a critical role in normal stem cell homing, also appear to be paramount to the homing of AML cells to, or retention within, the bone marrow.[77]

CD97 is differentially expressed in murine hematopoietic stem and progenitor cells (HSPCs),[78] but nothing is known about its expression in human hematopoietic progenitor cells. Wobus hypothesizes that CD97 is involved in AML progression and manifestation, potentially by interaction with its recently described ligand CD90/Thy-1, which is expressed by nonhematopoietic cells in the BM microenvironment.[79] They therefore initiated a comprehensive investigation of *de novo* AML samples and correlated the CD97 expression to clinically important parameters, such as NPM1 and FLT3 mutations.[80] The AML cell lines MV4–11 and EOL-1, as well as CD34[+] HSPCs, were used to study CD97 expression and regulation *in vitro*.

Compared to BM blasts of healthy donors, they detected significantly higher CD97 expression (mean fluorescence intensity, MFI) in 42% of AML samples. Patients with CD97 expression above the mean on leukemic blasts also showed increased expression of the molecule within the residual granulo- and monopoiesis. Of note, higher CD97 expression was accompanied by a significantly higher BM blast count (75% vs. 53%, $P < 0.001$). Interestingly, elevated CD97 expression was associated with mutations in NPM1 (46% vs. 18%, $P = 0.003$) and FLT3 genes (39% vs. 7%, $P < 0.001$), as well as lower CD34 expression (46% vs. 81%, $P < 0.001$). Furthermore, no AML1/ETO or CBFb/MYH11 fusion genes were detectable in CD97[+] AML versus 6% in CD97-AML.

Figure 12. (Top) Participants of the Adhesion-GPCR Workshop in front of the Residence Palace, Würzburg. Not shown: Demet Araç, Robert Kittel, Alexander Petrenko, Helgi Schiöth, Thue Schwartz, and Yuri Ushkaryov. (Bottom) Participants of the Adhesion-GPCR Workshop in the lecture hall of the Institute of Physiology, Würzburg. Not shown: Demet Araç, Davide Calebiro, Robert Kittel, Tobias Langenhan, Manja Wobus, and Lei Xu.

In vitro, Wobus detected lower CD97 expression levels in primary CD34$^+$ HSPC compared to the AML cell lines. Of note, in FLT3-ITD mutated MV4–11 cells, CD97 was expressed significantly higher. Treatment of this cell line with different tyrosine kinase inhibitors resulted in a decreased CD97 expression. The lower CD97 expression levels correlated with inhibition of the spontaneous migratory capacity. By using a dicer-substrate 27-mer duplex targeting CD97 in MV4–11 cells, Wobus knocked down expression to about 45%, which correlated with decreased transwell migration.

In summary, Wobus provides the first evidence of higher CD97 expression in AML cells compared to normal CD34$^+$ hematopoietic cells *in vivo* and *in vitro*, which correlates with FLT3-ITD mutation. In ongoing studies the underlying regulatory mechanisms will be investigated. The possible impact of CD97 as well as other molecules of that receptor family on AML biology and clinical outcome will be evaluated in a larger patient cohort.

Conclusions

The 6th International Adhesion-GPCR Workshop (photographs of workshop participants shown in Fig. 12) has yielded intricate details regarding the molecular faculty of these peculiar membrane proteins. The collective efforts of all the labs now allow us for the first time to speculate on the events from stimulus reception by an adhesion-GPCR via transduction to signal generation inside the cell, which is a remarkable milestone in the history of research on adhesion-GPCRs. The biological setting, the unusual important signaling route, became defined more clearly, allowing novel glimpses of adhesion-GPCR function and dysfunction. It is clear that interdisciplinarity and open scientific exchange is an important prerequisite and driving force to achieving this current level of understanding.

Acknowledgments

All participants of the 6th Adhesion-GPCR Workshop are grateful to Nicole Hartmann and Jennifer Gehring for their excellent organizational skills, and to the Interdisciplinary Center for Clinical Research (IZKF) of the University of Würzburg for its generous contribution toward the accommodation of workshop guests. We also acknowledge all colleagues that presented their data on adhesion-GPCRs during the poster session.

Conflicts of interest

The authors declare no conflicts of interest.

References

1. Yona, S., H.-H. Lin, W.O. Siu, *et al.* 2008. Adhesion-GPCRs: emerging roles for novel receptors. *Trends Biochem. Sci.* **33:** 491–500.
2. Kwakkenbos, M.J., E.N. Kop, M. Stacey, *et al.* 2004. The EGF-TM7 family: a postgenomic view. *Immunogenetics* **55:** 655–666.
3. Nordström, K.J.V., M. Sällman Almén, M.M. Edstam, *et al.* 2011. Independent HHsearch, Needleman–Wunsch-based, and motif analyses reveal the overall hierarchy for most of the G protein-coupled receptor families. *Mol. Biol. Evol.* **28:** 2471–2480.
4. Nordström, K.J.V., R. Fredriksson & H.B. Schiöth. 2008. The amphioxus (*Branchiostoma floridae*) genome contains a highly diversified set of G protein-coupled receptors. *BMC Evol. Biol.* **8:** 9.
5. Kamesh, N., G.K. Aradhyam & N. Manoj. 2008. The repertoire of G protein-coupled receptors in the sea squirt Ciona intestinalis. *BMC Biol. Biol.* **8:** 129.
6. Whittaker, C.A., K.-F. Bergeron, J. Whittle, *et al.* 2006. The echinoderm adhesome. *Dev. Biol.* **300:** 252–266.
7. Nordström, K.J.V., M.C. Lagerström, L.M.J. Wallér, *et al.* 2009. The secretin GPCRs descended from the family of adhesion GPCRs. *Mol. Biol. Evol.* **26:** 71–84.
8. Krishnan, A., M.S. Almén, R. Fredriksson & H.B. Schiöth. 2012. The origin of GPCRs: identification of mammalian like rhodopsin, adhesion, glutamate and frizzled GPCRs in fungi. *PLoS ONE* **7:** e29817.
9. Krasnoperov, V., M.A. Bittner, R.W. Holz, *et al.* 1999. Structural requirements for alpha-latrotoxin binding and alpha-latrotoxin-stimulated secretion. A study with calcium-independent receptor of alpha-latrotoxin (CIRL) deletion mutants. *J. Biol. Chem.* **274:** 3590–3596.
10. Lin, H.-H., G.-W. Chang, J.Q. Davies, *et al.* 2004. Autocatalytic cleavage of the EMR2 receptor occurs at a conserved G protein-coupled receptor proteolytic site motif. *J. Biol. Chem.* **279:** 31823–31832.
11. Volynski, K.E., J.-P. Silva, V.G. Lelianova, *et al.* 2004. Latrophilin fragments behave as independent proteins that associate and signal on binding of LTX(N4C). *EMBO J.* **23:** 4423–4433.
12. Araç, D., A.A. Boucard, M.F. Bolliger, *et al.* 2012. A novel evolutionarily conserved domain of cell-adhesion GPCRs mediates autoproteolysis. *EMBO J.* **31:** 1364–1378.
13. Paavola, K.J., J.R. Stephenson, S.L. Ritter, *et al.* 2011. The N terminus of the adhesion G protein-coupled receptor GPR56 controls receptor signaling activity. *J. Biol. Chem.* **286:** 28914–28921.
14. Krasnoperov, V.G., R. Beavis, O.G. Chepurny, *et al.* 1996. The calcium-independent receptor of alpha-latrotoxin is not a neurexin. *Biochem. Biophys. Res. Commun.* **227:** 868–875.
15. Krasnoperov, V.G., M.A. Bittner, R. Beavis, *et al.* 1997. alpha-Latrotoxin stimulates exocytosis by the interaction with a neuronal G-protein-coupled receptor. *Neuron* **18:** 925–937.

16. Langenhan, T., S. Prömel, L. Mestek, *et al.* 2009. Latrophilin signaling links anterior-posterior tissue polarity and oriented cell divisions in the C. elegans embryo. *Dev. Cell* **17:** 494–504.

17. Prömel, S., M. Frickenhaus, S. Hughes, *et al.* 2012. The GPS motif is a molecular switch for bimodal activities of adhesion class G protein-coupled receptors. *Cell Reports* **2:** 321–331.

18. Okajima, D., G. Kudo & H. Yokota. 2010. Brain-specific angiogenesis inhibitor 2 (BAI2) may be activated by proteolytic processing. *J. Recept. Signal Transduct. Res.* **30:** 143–153.

19. Silva, J.-P., V. Lelianova, C. Hopkins, *et al.* 2009. Functional cross-interaction of the fragments produced by the cleavage of distinct adhesion G-protein-coupled receptors. *J. Biol. Chem.* **284:** 6495–6506.

20. Iguchi, T., K. Sakata, K. Yoshizaki, *et al.* 2008. Orphan G protein-coupled receptor GPR56 regulates neural progenitor cell migration via a G alpha 12/13 and Rho pathway. *J. Biol. Chem.* **283:** 14469–14478.

21. Park, D., A.-C. Tosello-Trampont, M.R. Elliott, *et al.* 2007. BAI1 is an engulfment receptor for apoptotic cells upstream of the ELMO/Dock180/Rac module. *Nature* **450:** 430–434.

22. Bohnekamp, J. & T. Schöneberg. 2011. Cell adhesion receptor GPR133 couples to Gs protein. *J. Biol. Chem.* **286:** 41912–41916.

23. Gupte, J., G. Swaminath, J. Danao, *et al.* 2012. Signaling property study of adhesion G-protein-coupled receptors. *FEBS Lett.* **586:** 1214–1219.

24. Engelstoft, M.S., K.L. Egerod, B. Holst & T.W. Schwartz. 2008. A gut feeling for obesity: 7TM sensors on enteroendocrine cells. *Cell Metab.* **8:** 447–449.

25. Hansen, H.S., M.M. Rosenkilde, J.J. Holst & T.W. Schwartz. 2012. GPR119 as a fat sensor. *Trends Pharmacol. Sci.* **33:** 374–381.

26. Egerod, K.L., M.S. Engelstoft, K.V. Grunddal, *et al.* A major lineage of enteroendocrine cells co-express CCK, GLP-1, GIP, PYY, neurotensin, and secretin but not somatostatin. *Endocrinology.* In press.

27. Yang, L., G. Chen, S. Mohanty, *et al.* 2011. GPR56 regulates VEGF production and angiogenesis during melanoma progression. *Cancer Res.* **71:** 5558–5568.

28. Hamann, J., B. Vogel, G.M. van Schijndel & R.A. van Lier. 1996. The seven-span transmembrane receptor CD97 has a cellular ligand (CD55, DAF). *J. Exp. Med.* **184:** 1185–1189.

29. Veninga, H., R.M. Hoek, A.F. de Vos, *et al.* 2011. A novel role for CD55 in granulocyte homeostasis and anti-bacterial host defense. *PLoS ONE* **6:** e24431.

30. Hoek, R.M., D. de Launay, E.N. Kop, *et al.* 2010. Deletion of either CD55 or CD97 ameliorates arthritis in mouse models. *Arthritis. Rheum.* **62:** 1036–1042.

31. Ward, Y., R. Lake, J.J. Yin, *et al.* 2011. LPA receptor heterodimerizes with CD97 to amplify LPA-initiated RHO-dependent signaling and invasion in prostate cancer cells. *Cancer Res.* **71:** 7301–7311.

32. Chen, T.-Y., T.-L. Hwang, C.-Y. Lin, *et al.* 2011. EMR2 receptor ligation modulates cytokine secretion profiles and cell survival of lipopolysaccharide-treated neutrophils. *Chang Gung Med. J.* **34:** 468–477.

33. Yona, S., H.-H. Lin, P. Dri, *et al.* 2008. Ligation of the adhesion-GPCR EMR2 regulates human neutrophil function. *FASEB J.* **22:** 741–751.

34. Lin, H.-H., M. Stacey, S. Yona & G.-W. Chang. 2010. GPS proteolytic cleavage of adhesion-GPCRs. *Adv. Exp. Med. Biol.* **706:** 49–58.

35. Huang, Y.-S., N.-Y. Chiang, C.-H. Hu, *et al.* Lin. 2012. Activation of myeloid cell-specific adhesion class G protein-coupled receptor EMR2 via ligation-induced translocation and interaction of receptor subunits in lipid raft microdomains. *Mol. Cell Biol.* **32:** 1408–1420.

36. Lohse, M.J., V.O. Nikolaev, P. Hein, *et al.* 2008. Optical techniques to analyze real-time activation and signaling of G-protein-coupled receptors. *Trends Pharmacol. Sci.* **29:** 159–165.

37. Calebiro, D., V.O. Nikolaev, M.C. Gagliani, *et al.* 2009. Persistent cAMP-signals triggered by internalized G-protein-coupled receptors. *PLoS Biol.* **7:** e1000172.

38. Nikolaev, V.O., M. Bünemann, L. Hein, *et al.* 2004. Novel single chain cAMP sensors for receptor-induced signal propagation. *J. Biol. Chem.* **279:** 37215–37218.

39. Calebiro, D., V.O. Nikolaev, L. Persani & M.J. Lohse. 2010. Signaling by internalized G-protein-coupled receptors. *Trends Pharmacol. Sci.* **31:** 221–228.

40. Monk, K.R., S.G. Naylor, T.D. Glenn, *et al.* 2009. A G protein-coupled receptor is essential for Schwann cells to initiate myelination. *Science* **325:** 1402–1405.

41. Monk, K.R., K. Oshima, S. Jörs, *et al.* 2011. Gpr126 is essential for peripheral nerve development and myelination in mammals. *Development* **138:** 2673–2680.

42. Waller-Evans, H., S. Prömel, T. Langenhan, *et al.* 2010. The orphan adhesion-GPCR GPR126 is required for embryonic development in the mouse. *PLoS ONE* **5:** e14047.

43. Stehlik, C., R. Kroismayr, A. Dorfleutner, *et al.* 2004. VIGR—a novel inducible adhesion family G-protein coupled receptor in endothelial cells. *FEBS Lett.* **569:** 149–155.

44. Moriguchi, T., K. Haraguchi, N. Ueda, *et al.* 2004. DREG, a developmentally regulated G protein-coupled receptor containing two conserved proteolytic cleavage sites. *Genes Cells* **9:** 549–560.

45. Tissir, F., I. Bar, Y. Jossin, *et al.* 2005. Protocadherin Celsr3 is crucial in axonal tract development. *Nat. Neurosci.* **8:** 451–457.

46. Tissir, F., Y. Qu, M. Montcouquiol, *et al.* 2010. Lack of cadherins Celsr2 and Celsr3 impairs ependymal ciliogenesis, leading to fatal hydrocephalus. *Nat. Neurosci.* **13:** 700–707.

47. Qu, Y., D.M. Glasco, L. Zhou, *et al.* 2010. Atypical cadherins Celsr1–3 differentially regulate migration of facial branchiomotor neurons in mice. *J. Neurosci.* **30:** 9392–9401.

48. Song, H., J. Hu, W. Chen, *et al.* 2010. Planar cell polarity breaks bilateral symmetry by controlling ciliary positioning. *Nature* **466:** 378–382

49. Curtin, J.A., E. Quint, V. Tsipouri, *et al.* 2003. Mutation of Celsr1 disrupts planar polarity of inner ear hair cells and causes severe neural tube defects in the mouse. *Curr. Biol.* **13:** 1129–1133.

50. Yates, L.L., C. Schnatwinkel, J.N. Murdoch, *et al.* 2010. The PCP genes Celsr1 and Vangl2 are required for normal lung branching morphogenesis. *Hum Mol Genet* **19:** 2251–2267.

51. Devenport, D., D. Oristian, E. Heller & E. Fuchs. 2011. Mitotic internalization of planar cell polarity proteins preserves tissue polarity. *Nat. Cell Biol.* **13:** 893–902.

52. Formstone, C.J., C. Moxon, J. Murdoch, *et al.* 2010. Basal enrichment within neuroepithelia suggests novel function(s) for Celsr1 protein. *Mol. Cell. Neurosci.* **44:** 210–222.

53. Yates, L.L., C. Schnatwinkel, L. Hazelwood, *et al.* Scribble is required for normal epithelial cell-cell contacts and lumen morphogenesis in the mammalian lung. *Dev Biol.* In press.

54. Becker, S., E. Wandel, M. Wobus, *et al.* 2010. Overexpression of CD97 in intestinal epithelial cells of transgenic mice attenuates colitis by strengthening adherens junctions. *PLoS ONE* **5:** e8507.

55. Cummins, A.G. & F.M. Thompson. 2002. Effect of breast milk and weaning on epithelial growth of the small intestine in humans. *Gut* **51:** 748–754.

56. Tobaben, S., T.C. Südhof & B. Stahl. 2002. Genetic analysis of alpha-latrotoxin receptors reveals functional interdependence of CIRL/latrophilin 1 and neurexin 1 alpha. *J. Biol. Chem.* **277:** 6359–6365.

57. Davletov, B.A., O.G. Shamotienko, V.G. Lelianova, *et al.* 1996. Isolation and biochemical characterization of a Ca2+-independent alpha-latrotoxin-binding protein. *J. Biol. Chem.* **271:** 23239–23245.

58. Davletov, B.A., F.A. Meunier, A.C. Ashton, *et al.* 1998. Vesicle exocytosis stimulated by alpha-latrotoxin is mediated by latrophilin and requires both external and stored Ca2. *EMBO J.* **17:** 3909–3920.

59. Silva, J.-P. & Y.A. Ushkaryov. 2010. The latrophilins, "split-personality" receptors. *Adv. Exp. Med. Biol.* **706:** 59–75.

60. Silva, J.-P., V.G. Lelianova, Y.S. Ermolyuk, *et al.* 2011. Latrophilin 1 and its endogenous ligand Lasso/teneurin-2 form a high-affinity transsynaptic receptor pair with signaling capabilities. *Proc. Natl. Acad. Sci. USA* **108:** 12113–12118.

61. Silva, J.P., N. Vysokov, G.V. Lelianova, *et al.* Interaction between latrophilin-1 and Lasso is required for functional maturation of synapses. Manuscript in preparation.

62. McMillan, D.R. & P.C. White. 2010. Studies on the very large G protein-coupled receptor: from initial discovery to determining its role in sensorineural deafness in higher animals. *Adv. Exp. Med. Biol.* **706:** 76–86.

63. McGee, J., R.J. Goodyear, D.R. McMillan, *et al.* 2006. The very large G-protein-coupled receptor VLGR1: a component of the ankle link complex required for the normal development of auditory hair bundles. *J. Neurosci.* **26:** 6543–6553.

64. Maerker, T., E. van Wijk, N. Overlack, *et al.* 2008. A novel Usher protein network at the periciliary reloading point between molecular transport machineries in vertebrate photoreceptor cells. *Hum. Mol. Genet.* **17:** 71–86.

65. Wolfrum, U. 2011. Protein networks related to the Usher syndrome gain insights in the molecular basis of the disease. In *Usher Syndrome: Pathogenesis, Diagnosis and Therapy.* S.A. Ed.: 51–73. Nova Science Publishers, Inc. Hauppauge, NY.

66. Reiners, J., E. van Wijk, T. Märker, *et al.* 2005. Scaffold protein harmonin (USH1C) provides molecular links between Usher syndrome type 1 and type 2. *Hum. Mol. Genet.* **14:** 3933–3943.

67. Piao, X., B.S. Chang, A. Bodell, *et al.* 2005. Genotype-phenotype analysis of human frontoparietal polymicrogyria syndromes. *Ann. Neurol.* **58:** 680–687.

68. Piao, X., R.S. Hill, A. Bodell, *et al.* 2004. G protein-coupled receptor-dependent development of human frontal cortex. *Science* **303:** 2033–2036.

69. Luo, R., S.-J. Jeong, Z. Jin, *et al.* 2011. G protein-coupled receptor 56 and collagen III, a receptor-ligand pair, regulates cortical development and lamination. *Proc. Natl. Acad. Sci. USA* **108:** 12925–12930.

70. Li, S., Z. Jin, S. Koirala, *et al.* 2008. GPR56 regulates pial basement membrane integrity and cortical lamination. *J. Neurosci.* **28:** 5817–5826.

71. Jeong, S.-J., S. Li, R. Luo, *et al.* 2012. Loss of Col3a1, the gene for Ehlers-Danlos syndrome type IV, results in neocortical dyslamination. *PLoS ONE* **7:** e29767.

72. Jeong, S.-J., R. Luo, S. Li, *et al.* 2012. Characterization of G protein-coupled receptor 56 protein expression in the mouse developing neocortex. *J. Comp. Neurol.* **520:** 2930–2940.

73. Xu, L., S. Begum, J.D. Hearn & R.O. Hynes. 2006. GPR56, an atypical G protein-coupled receptor, binds tissue transglutaminase, TG2, and inhibits melanoma tumor growth and metastasis. *Proc. Natl. Acad. Sci. USA* **103:** 9023–9028.

74. Lorand, L. & R.M. Graham. 2003. Transglutaminases: crosslinking enzymes with pleiotropic functions. *Nat. Rev. Mol. Cell Biol.* **4:** 140–156.

75. Belkin, A.M. 2011. Extracellular TG2: emerging functions and regulation. *FEBS J.* **278:** 4704–4716.

76. Davies, J.Q., H.-H. Lin, M. Stacey, *et al.* 2011. Leukocyte adhesion-GPCR EMR2 is aberrantly expressed in human breast carcinomas and is associated with patient survival. *Oncol. Rep.* **25:** 619–627.

77. Becker, P.S. 2012. Dependence of acute myeloid leukemia on adhesion within the bone marrow microenvironment. *The Scientific World Journal* **2012:** 856467.

78. van Pel, M., H. Hagoort, J. Hamann & W.E. Fibbe. 2008. CD97 is differentially expressed on murine hematopoietic stem-and progenitor-cells. *Haematologica* **93:** 1137–1144.

79. Wandel, E., A. Saalbach, D. Sittig, *et al.* 2012. Thy-1 (CD90) is an interacting partner for CD97 on activated endothelial cells. *J. Immunol.* **188:** 1442–1450.

80. Kiyoi, H. & T. Naoe. 2006. Biology, clinical relevance, and molecularly targeted therapy in acute leukemia with FLT3 mutation. *Int. J. Hematol.* **83:** 301–308.

Appendix

6th International Adhesion-GPCR Workshop

6–8 September 2012 – Würzburg – Germany

Scientific program

Thursday, Sep 6 2012

9:00 – 9:10
Opening remarks
Tobias Langenhan, University of Würzburg

Session A – Structural hallmarks of Adhesion-GPCR

9:10 – 9:35
  The GPS motif: 15 years of studies
  Alexander Petrenko, Russian Academy of Sciences, Moscow

9:35 – 10:00
  A Novel Evolutionarily Conserved Domain of Cell-Adhesion GPCRs Mediates Autoproteolysis
  Demet Araç-Ozkan, Stanford University

10:00 – 10:25
  Structural insights into the adhesion-GPCR CD97
  Martin Stacey, University of Leeds

10:25 – 10:45
  Coffee break

Session B – Neurobiological roles of Adhesion-GPCR

10:45 – 11:10
  Latrophilin receptors regulate presynaptic transmitter release
  Tobias Langenhan, University of Würzburg

11:10 – 11:35
  High-affinity functional trans-synaptic receptor pairs between presynaptic latrophilin and postsynaptic Lasso (teneurin-2)
  Yuri Ushkaryov, University of Kent

11:35 – 12:00
  GPR56-dependent development of the frontal cerebral cortex
  Xianhua Piao, Harvard Medical School

12:00 – 14:00
  Lunch break

Session C – Neurobiological roles of Adhesion-GPCR (continued)

14:00 – 14:25
  GPR56, together with $\alpha 3 \beta 1$ Integrin, Regulates Cortical Lamination
  Kathleen Singer, Harvard Medical School

14:25 – 14:50
  Role of Celsr1–3 cadherins in planar cell polarity and brain development
  André Goffinet, University of Louvain

14:50 – 15:15
  The very large G protein coupled receptor Vlgr1b/GPR98 as a key component of the

Usher syndrome protein networks in the inner ear and the retina
  Uwe Wolfrum, University of Mainz

15:15 – 15:40
  Molecular and genetic analysis of Gpr126 in peripheral nerve development
  Kelly Monk, Washington University School of Medicine, St. Louis

15:40 – 16:30
  Coffee break

19:00
  Evening program

Friday, Sep 7 2012

Session D – Adhesion-GPCR in development

9:00 – 9:25
  Basal enrichment of Celsr1 protein within epithelia: novel function or apico-basal dependent planar cell polarity (PCP) signalling?
  Caroline Formstone, King's College London

9:25 – 9:50
  Knockdown of the orphan G protein-coupled receptor 126 influences ventricular morphogenesis and heart function in zebrafish and mice
  Felix Engel, Max-Planck-Institute Bad Nauheim

9:50 – 10:15
  Mice constitutively overexpressing CD97 in enterocytes develop a megaintestine without alterations in histology and cell fate decision
  Gabriela Aust, University of Leipzig

10:15 – 10:45
  Coffee break

Session E – Adhesion-GPCR in tumor biology

10:45 – 11:10
  Roles of GPR56 and TG2 during Melanoma Progression
  Lei Xu, Rochester School of Medicine

11:10 – 11:35
  Molecular characterization of the interaction of GPR56 and a novel ligand
  Hsi-Hsien Lin, Chang Gung University

11:35 – 12:00
  The expression of the EGF-TM7 receptor CD97 is higher in CD34-negative and NPM1/FLT3-ITD mutated AML
  Manja Wobus, University of Dresden

12:00 – 14:00
   Lunch break

Session F – Mechanisms of signal transduction of Adhesion-GPCR

14:00 – 14:25
   Shear stress-dependent downregulation of the adhesion-GPCR CD97 on circulating leukocytes upon contact with its ligand CD55
   Jörg Hamann, University of Amsterdam

14:25 – 14:50
   Insights into the molecular function of latrophilins – logic of adhesion-GPCR signalling
   Simone Prömel, University of Oxford & University of Leipzig

14:50 – 15:15
   G protein-mediated signal transduction of adhesion GPCR
   Ines Liebscher, University of Leipzig

15:15 – 15:40
   Real-time monitoring of GPCR signaling in living cells: from intracellular signaling microdomains to single molecules
   Davide Calebiro, University of Würzburg

15:40 – 16:30
   Coffee break

16:30 – 17:30
   General Meeting of the Adhesion-GPCR Consortium

19:00
   Evening program

Saturday, Sep 8 2012

Session G – Adhesion-GPCR in endocrine, cardiovascular, and immune functions

9:00 – 9:25
   The origin of the Adhesion-GPCR family
   Helgi Schiöth, University of Uppsala

9:25 – 9:50
   The ADHD-susceptibility gene *lphn3.1* modulates dopaminergic neuron formation and locomotor activity during zebrafish development
   Klaus-Peter Lesch, University of Würzburg

9:50 – 10:15
   Adhesion 7TM receptors – major players in the endocrine and enteroendocrine system
   Thue Schwartz, University of Copenhagen

10:15 – 10:45
   Coffee break

10:45 – 11:45
   Future initiatives of the Adhesion-GPCR community

11:45 – 12:00
   Concluding remarks
   Jörg Hamann, University of Amsterdam & Tobias Langenhan, University of Würzburg

Ann. N.Y. Acad. Sci. ISSN 0077-8923

ANNALS OF THE NEW YORK ACADEMY OF SCIENCES

Issue: Annals *Meeting Reports*

Scientific considerations for complex drugs in light of established and emerging regulatory guidance

Chris Holloway,[1] Jan Mueller-Berghaus,[2] Beatriz Silva Lima,[3] Sau (Larry) Lee,[4] Janet S. Wyatt,[5] J. Michael Nicholas,[6] and Daan J.A. Crommelin[7]

[1]ERA Consulting Group, Walsrode, Germany. [2]Paul-Ehrlich-Institut, Federal Institute for Vaccines and Biomedicines, Langen, Germany. [3]University of Lisbon, Lisbon, Portugal. [4]U. S. Food and Drug Administration, Silver Spring, Maryland. [5]Institute of Pediatric Nursing, Gaithersburg, Maryland. [6]Teva Pharmaceuticals, Petach Tikva, Israel. [7]Utrecht University, Utrecht, the Netherlands

Address for correspondence: annals@nyas.org

On March 9, 2012, the New York Academy of Sciences brought together experts representing a variety of perspectives—including academic, industrial, regulatory, as well as those from physicians and consumers—to discuss considerations for the non-biological complex drug (NBCD) regulatory approval pathway, given the emerging regulatory guidelines for biosimilars (follow-on biological complex drugs). Some of the organizers of the conference expressed their belief that NBCDs share a number of characteristic features with biologicals: the structure cannot be fully defined by the available (physicochemical) analytical tests, and quality assurance is based on in-depth knowledge, consistency, and control of the production process. However, their view on NBCDs was not universally accepted among the experts who participated in the conference. Plenary sessions addressed the most recent regulatory developments, experimental design, interchangeability, and immunogenicity issues for follow-on versions of complex drugs from the perspective of key audiences, including industry, regulatory agencies, physicians, and consumers. This report summarizes these various perspectives on NBCDs and the scientific and regulatory considerations associated with complex drug categories.

Keywords: biosimilars; follow-on biologics; complex drug; NBCD

Background and context: keynote address

Session Chairs: J. Michael Nicholas and Daan J.A. Crommelin

Since 2009, the U.S. Food and Drug Administration (FDA) has been working toward implementing the Biologics Price Competition and Innovation Act in order to establish an abbreviated approval pathway for biological products that are demonstrated to be "highly similar" to, or "interchangeable" with, previously approved and regulated drugs. While the much anticipated draft of the FDA guidelines for

biosimilars was released on February 9, 2012, current U.S. regulations do not make scientific distinctions between small-molecule drugs and non-biological complex drugs (NBCDs), although the latter may present many of the same scientific and clinical challenges to reproduce as biologics.

The FDA guidelines released in February 2012 are an important step in bringing affordable versions of complex drugs to market. As regulatory authorities continue to provide direction in follow-on applications, according to the session chairs, distinctions should be made between biologics and NBCDs in order to provide a guide for industry to use when submitting applications.

The New York Academy of Sciences' conference, "Scientific Considerations for Complex Drugs in Light of Established Regulatory Guidance," convened with a keynote address that provided an overview of the scientific and regulatory

---

The opinions expressed in this article are those of the authors and not of any regulatory, industry, or medical organization they may be affiliated with, including the European Medicines Agency (EMA) and the U. S. Food and Drug Administration (FDA).

doi: 10.1111/j.1749-6632.2012.06811.x

considerations of small molecules, biologics, and NBCDs from different perspectives.

The scientific and regulatory differences between small molecules, biologics, and non-biological complex drugs

In his keynote address, Chris Holloway (ERA Consulting Group) noted that medicinal products can broadly, but conveniently, be divided into three classes: small molecules, biologics, and NBCDs. These drug classes differ in their regulatory requirements, especially when considering follow-on versions of these products. Generally, follow-on products containing a small molecular entity as an active substance are eligible for approval as generics, whereas biologics are not. The scientifically justified paradigm for biological medicinal products, anchored in U.S. and EU regulations and those of other developed jurisdictions, is that such products are inexorably linked to their specific manufacturing processes and that physicochemical characterization and demonstration of bioequivalence will not suffice as evidence of efficacy and safety.

Holloway exemplified potential issues associated with biological manufacturing processes through the case of eosinophilia-myalgia syndrome (EMS), which was linked to a tryptophan product in the late 1980s.[1] Tryptophan was commonly used as a supplement for sleep induction in the United States. Showa Denko, the major supplier of tryptophan at that time, genetically modified the bacterial production strain for increased productivity and also amended the purification process. The emergence of several new impurities went unnoticed, and these impurities were implicated in as many as 60,000 cases of EMS in the United States, of which at least 1500 resulted in severe debilitation, with at least 37 deaths.

The major concern associated with therapeutics derived from recombinant DNA technology is their potential immunogenicity, which cannot be predicted from physicochemical analysis. Wadhwa *et al.* showed that two granulocyte-macrophage colony-stimulating factors (GM-CSFs) exhibited very different immunogenicity profiles in metastatic colorectal cancer patients on GM-CSF combination therapy.[2] The clinical consequences of immunogenicity are well illustrated by cases of pure red cell aplasia. This was caused by neutralizing antibodies to a recombinant epoetin product in the late 1990s, believed to have been associated with syringe plunger-stopper leachates acting as weak adjuvants.[3]

Patent expiry on several highly successful biologics has spawned the development of follow-on versions, commonly known as "biosimilars." By 2006, the legal and regulatory basis for marketing authorization of such products had been established in Europe,[4] followed more recently by the corresponding legislation in the United States.[5] In Europe, several biosimilar growth hormones, epoetins, and filgrastim products have already been approved through this route. However, successful development of a biosimilar product is neither trivial, nor a foregone conclusion, and several products have failed the regulatory process. We can learn as much from these failures as from the successes. Alpheon® (interferon-α-2a), a product claimed to be biosimilar to Roferon-A® (interferon-α-2a, recombinant), was found to have a different impurity profile from the reference product. In a clinical trial, more hepatitis C patients experienced a return of the disease after treatment with Alpheon was stopped than with Roferon, and there were more side effects with Alpheon.[6] Perhaps this was due to greater immunogenicity of Alpheon, but regulators were unable to assess this, as the immunogenicity test was not sufficiently validated. One would expect a biosimilar insulin to be more straightforward, but Marvel's insulin products were not approved when first submitted, because studies failed to demonstrate bioequivalence, and a comparative phase III trial appeared to favor the reference product Humulin® over Marvel's products.[7]

More recently, the marketing authorization application for an epoetin (Epostim®, GeneMedix PLC) was withdrawn by the applicant after the Committee for Medicinal Products for Human Use (CHMP) issued its assessment report and list of questions and objections.[8] These case studies show that compliance with requirements for biosimilarity is not a foregone conclusion, and that development of a biosimilar product is not trivial. Further, such case studies indicate the appropriateness of the regulatory approach for biosimilars, in that appropriate, comparative clinical trials are needed for purposes of distinguishing an "inferior" follow-on product.

Although not formally submitted as a biosimilar application, a follow-on interferon-β-1a, Biferonex®, depended, for its hybrid approach, on data previously generated with the reference product Avonex®. The application was withdrawn after the regulators expressed concerns relating to deamidation, lack of assurance on comparability, and questionable clinical efficacy. The assessors were unable to determine whether clinical failure was due to study design, robustness of the results, or intrinsic product properties.[9] These examples serve to emphasize the importance of well-designed, comparative clinical trials to support efficacy and safety.

In Holloway's opinion, the plethora of EU guidelines on biosimilar products provides sound guidance for the development of these products, which can usefully be augmented through productive interactions (scientific advice) with regulators. By contrast, the regulatory basis for NBCDs is far less clear and guidelines are largely not available. However, the development of such guidelines could be important, as illustrated by the following two examples of NBCDs.

The iron-sucrose complex (ISC) is a medicinal product composed of polynuclear iron (III)-oxyhydroxide coated with a carbohydrate ligand. It is administered intravenously in patients with iron deficiency, notably in chronic kidney disease, to replenish body iron stores. In a reflection paper, the European Medicines Agency (EMA) noted that physicochemical characteristics and pharmacokinetics may not suffice to ensure safety and efficacy of a follow-on ISC product.[10] Clark, reporting results from Toblli *et al.*,[11] noted different effects of European iron sucrose–similar (ISS) preparations and originator iron sucrose (ISO) on nitrosative stress, apoptosis, oxidative stress, and biochemical and inflammatory markers in rats. Differences were also noted in clinical practice. When an ISS was introduced in place of an ISO in a dialysis unit in France, shifts in hemoglobin and iron indices in the patients were observed; it was concluded that an ISS was not therapeutically equivalent to an ISO.[12]

Glatiramer acetate (GA) belongs to the class of glatiramoids and is the active substance in Copaxone®, which, alongside interferon-β, is a mainstay product in the treatment of relapsing-remitting multiple sclerosis. GA is a complex, synthetic polypeptide containing the amino acids L-glutamic acid, L-alanine, L-lysine, and L-tyrosine, with the molecular formula (L-Glu13–15, L-Ala39–46, L-Tyr8.6–10, L-Lys30–37)n mX. The amino acid sequences are not totally random, but the actual sequences are known to depend heavily on the manufacturing process. The complexity of GA is exacerbated by the fact that the exact mechanism of action is unknown, and the epitopes responsible for efficacy and safety cannot be identified. Thus, it is impossible to prove, on the basis of characterization alone, that a follow-on glatiramoid is comparable to Copaxone. Indeed, in a reflection paper on nanotechnology-based products, the EMA mentioned Copaxone in this context,[13] thereby underlining the complexity of this NBCD.

While a follow-on glatiramoid may appear to be similar using typical release standards for this class of compound, in-depth characterization can reveal differences. For example, methods to examine higher-order structure have revealed subtle differences between Copaxone and certain follow-on glatiramoids, the clinical consequences of which are unknown. However, increased cross-linking is expected to yield degradation products *in vivo* that are not present in Copaxone, which may have safety implications. In contrast to biosimilars, for which long-term toxicology studies in animals are not considered relevant, such studies may be important for glatiramoids. The development some years ago of a second-generation, more potent glatiramoid, TV-5010, showed no issues in short-term toxicity studies, but marked toxicity was noted in chronic toxicology studies ≥ six months. Clinically, TV-5010 was not equivalent to Copaxone, as the profile of antibody classes generated over time appeared to be different.

These CMC (chemistry, manufacturing, and controls), nonclinical and clinical experiences highlight the need for a clear understanding of the difficult issues around development of follow-on glatiramoids. In Holloway's opinion, this is a clear case for which clarification of a regulatory pathway and appropriate guidelines on requirements for product development and approval would be beneficial to ensure that any follow-on product is efficacious and safe. In contrast to biosimilar interferon-β products, for which a draft guideline has been issued in Europe,[14] the approach for follow-on glatiramoids will need to apply a different paradigm, according to Holloway. First, for follow-on interferon-β products, a repeat-dose toxicity study is generally not required. The TV-5010 experience demonstrates the importance of long-term toxicity studies for glatiramoids. Moreover, the biosimilar interferon-β guideline does not require that sponsors demonstrate clinical benefit per se, as this has already been established with the reference product. Glatiramoids are less amenable to a CMC-based similarity exercise than an interferon-β, which has a clear and unique amino acid sequence. Therefore, demonstration of clinical benefit should surely be required for follow-on glatiramoids, given that the species within the complex mixture that are responsible for efficacy and, for that matter, safety, are unknown.

Biosimilar monoclonal antibodies: current European developments

Jan Mueller-Berghaus (Paul-Ehrlich-Institut, Federal Institute for Vaccines and Biomedicines) introduced European developments with biosimilar monoclonal antibodies (mABs). More than 10 years ago, European legislators implemented a regulatory pathway for the approval of biosimilars. The law specifically stipulates that "…results of appropriate pre-clinical tests or clinical trials" must be provided.[15] With the help of its working parties (e.g., the Scientific Advice Working Party, the Biosimilar Medicinal Products Working Party, the Safety Working Party, the Biologics Working Party, and the Biostatistics Working Party), the CHMP at the EMA has published several guidelines over the years relating to general principles of biosimilar philosophy and development, as well as product-specific guidelines for biosimilar insulin, somatropin, G-CSF, and epoetin, for example.

According to Mueller-Berghaus, the therapeutic importance and commercial success of mABs also make this product class an attractive target for biosimilar developers. However, it has to be recognized that mABs are, by far, the largest proteins considered for biosimilar developments to date. They have a complex structure and have demonstrated multifunctionality, including binding to antigens, binding to several distinct cellular receptors, as well as activating complement. In the past years, further complexity has been recognized by the findings that posttranslational modifications, especially in the Fc part of the molecules, can have a profound impact on specific functions.

Driven by scientific advice requests from biosimilar developers for mABs, CHMP has provided product-specific advice to a number of companies; however, it was recognized that general guidance could be helpful for developers. The first draft of "Guideline on similar biological medicinal products containing monoclonal antibodies—non-clinical and clinical issues" was published in November 2010 and was extensively commented on until May 2011. It became available in May 2012.[16]

Mueller-Berghaus noted that current analytical tools have ever-increasing sensitivity and thus improved ability to detect and describe the heterogeneity of a mAB product. Since the methods are applicable to any protein therapeutic, experts felt that a specific guideline for the pharmaceutical quality and characterization of the molecule would not be required.

Regarding the nonclinical evaluation of mABs, a stepwise approach is recommended. Comparative *in vitro* studies are necessary to evaluate the different functionalities of a mAB, which should include target binding, Fc receptor binding, complement binding, and subsequent functional characterization, using, for example, antibody-dependant cellular cytotoxicity and complement activation. Importantly, a comprehensive characterization is recommended, as there still are significant gaps in the knowledge of which function of an antibody drives clinical efficacy and safety. Only after finalization and assessment of *in vitro* studies does the need for nonclinical *in vivo* studies become clear. Toxicity studies in nonrelevant species and nonhuman primates are not generally recommended.

Several strategies for comparative pharmacokinetic testing can be applied and have to be determined on a case-by-case basis; the use of a single-dose study in healthy volunteers is preferred, but this study may often not be feasible.

The use of a dose-sensitive model for clinical comparison is encouraged. Unfortunately, there are few validated biomarkers that can be used, and the use of a suboptimal dose in patients is ethically not feasible. Normally, therefore, equivalence trials are expected as part of the evaluation of biosimilarity. For methodological reasons, a homogenous and sensitive population is recommended, and deviations from disease-specific regulatory guidelines may be necessary. In the case of several indications for one medicinal product, the extrapolation to other indications may be possible.

The legal framework in Europe appears to imply that only reference medicinal products sourced in Europe can be used for the demonstration of biosimilarity. However, the debate is ongoing, and interested parties are strongly advised to follow the discussion closely. In summary, a pathway for the approval of biosimilar monoclonal antibodies has been laid out in Europe, and the first submissions are expected soon.

Insights into the process of establishing regulations for NBCDs

According to Beatriz Silva Lima (University of Lisbon), examples of NBCDs currently in the market, or under development, are nano-sized formulations such as iron sucrose, liposome, transferosome, block copolymer micelles, and many others. Although not explicitly described under the definition, nano-sized medicines have been under the scrutiny of regulatory authorities at a global level. In the European Union, the EMA has put in place several initiatives, starting with the implementation of the EMA Ad-Hoc Nanomedicine Expert Group in 2009, which organized the International Workshop on Nanomedicines in 2010, together with international regulatory partners from the United States, Japan, and Canada. Currently at the EMA, there is a drafting group on nanomedicines. Experts in the areas of quality, nonclinical and clinical efficacy, and safety are preparing a set of reflection papers covering the development of several types of nanomedicines well within the scope of the definition of NBCDs. The need (or lack thereof) of additional regulatory guidance to support the development of NBCDs, based on nanoparticles (NP), may depend on the existing knowledge of the NP type, the active substance (AS), and the final complex drug (NP/AS). Different scenarios can be envisaged (Fig. 1).

Silva Lima noted that for NBCDs, including a novel active substance integrated in a known or an innovative (particle) system, the principles as stated in existing safety and efficacy nonclinical and clinical guidelines can be considered appropriate, and so a new general guidance may not be needed. In principle, the quality, efficacy, and safety of the new active substance, including the characterization of the NP entity and of the final NBCD, will need to be addressed as in any other development plan for a new medicinal product. In the case of NBCDs consisting of well-known active substances formulated with either well-known or new (nano)systems, there will most possibly be a need, determined on a case-by-

Regulatory Thinking on Nanosized Formulations

- New NP / New AS

- "Known" NP / New AS

- New NP / Known AS

- "Known" NP / Known AS

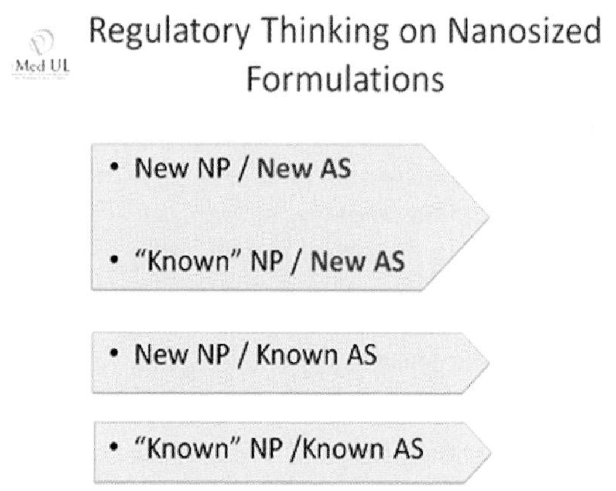

Figure 1. Regulatory thinking on nanosized formulations.

case basis, to further address the potential impact of the new complex on the previously defined efficacy and safety profile of the active substance, including its disposition.

Therefore, when integrating a known active substance in an NBCD, the efficacy and safety, dose–response relationships, and pharmacokinetic/pharmacodynamic relationships may need to be revisited and redefined. The proof of concept of the NBCD may need to be discussed again, since, in many cases, it may justify the purpose of the new formulation (e.g., different or selective tissue or cell targeting, or modified potency) on the basis of the type, structure, size, and other physical parameters of the particles and their components.

Establishing the comparability of two NBCDs to analyze their similarity involves—in relation to the comparison of candidate generic formulations with the test reference—a more complex process owing to the multiple variability factors associated with the particle component. Examples of comparability under regulatory discussions in the public domain include liposomal formulations of doxorubicin and iron glucose NP formulations. For both cases, it is recognized that, in contrast to noncomplex formulations of small chemical entities, comparative plasma concentrations of the active substances with liposomal or other NP formulations may not be sufficient to deduce similarity of efficacy and safety, since NPs may distribute differently in tissues and in cells at cellular and subcellular compartments. Additionally, similarity cannot be concluded solely on the basis of the levels of total drug reached in different compartments. Rather, a discrimination of free versus complexed forms will be needed. This often poses substantial analytical challenges. There is a concern that small differences in particle/complex attributes may relevantly change the activity of the active substance, its distribution profile, and/or its persistence at tissue/cellular or subcellular levels. This may have an impact on the efficacy as compared to the reference product and/or differences in the toxicological profile.[17] To address these concerns, contrary to classical generic drugs, researchers may need to use nonclinical studies to examine the comparability of NBCDs in order to address the aspects that cannot be studied clinically, for example, tissue/cellular biodistribution. However, scientists from industry and regulatory agenicies generally agree that comparative animal toxicology studies may not be sufficiently powered to identify differences in toxicological properties of two formulations. In order to achieve a reliable toxicological comparison between two formulations, the number of animals required might be too high, which poses practical and ethical difficulties. Most often the outcome of conducted studies will suggest nondissimilarity, leading to a false sense of reassurance, which is a safety concern. Therefore, if comparative animal studies are required, these might mainly address the comparative biodistribution of the similar candidate with the reference drug. The need to compare specific toxicological attributes, such as cardiotoxicity, should be case based.

While acknowledging the difficulties being raised regarding comparative animal studies, the way forward in NBCD comparability exercises may be toward integration of *in silico*, *in vitro*, (and *in vivo*) data regarding the particle/system attributes versus their biological behavior, in order to provide modeling strategies for systems concepts and for fixing the attributes determining the similarity to be established.

Silva Lima recognized that knowledge in the area of NBCDs is still growing concerning the relationship among particle/system attributes and, for example, biodistribution, accumulation, persistence, and activity. As discussed by MacNeil, the biocompatibility, tropism, and renal versus liver clearance profile, for example, can be predicted when integrating particle attributes like zeta potential, surface charge, size, and composition.[18] As for the future of mABs, the progress of science should provide tools for proof of similarity of these systems using mod-

ern techniques that focus on the quality attributes and *in vitro* profile as determinants of comparison ("systems sciences"). The reflection papers under development at the EMA on comparison of liposomes,[19] iron nanoparticles,[20] or block copolymers encourage these strategies.

Therapeutic equivalence of complex drug products: case study of generic calcitonin nasal spray products

Sau (Larry) Lee (FDA) discussed the FDA requirements for complex drugs, using generic calcitonin nasal spray products as an example. Under the Abbreviated New Drug Application (ANDA) pathway, a proposed generic salmon calcitonin nasal spray is required to demonstrate pharmaceutical equivalence and bioequivalence to the brand name counterpart or the reference listed drug (RLD). Pharmaceutical equivalence requires that the generic drug product contains the same active ingredient(s) as the RLD, that it be identical in strength, dosage form, and route of administration, and that it meet compendial or other applicable standards of strength, quality, purity, and identity.[21]

Bioequivalence refers to the absence of a statistically significant difference in the rate and extent to which the active ingredient in a pharmaceutically equivalent drug product becomes available at the site of action, when administered to subjects at the same molar dose under similar conditions.[21] Once a sponsor of an ANDA provides sufficient data to demonstrate that a generic drug product is pharmaceutically equivalent and bioequivalent to the RLD, the FDA deems these two drug products to be therapeutic equivalents. Therapeutically equivalent drug products are expected to have the same clinical efficacy and safety profiles when administered to patients under conditions specified in the labeling.

The Office of Generic Drugs of the FDA has reviewed and approved numerous generic complex drug products, such as synthetic peptide products, heparin, and low-molecular-weight heparin products. In comparison to conventional "small molecule" drugs, review and approval of complex drug products under the ANDA regulatory pathway is considerably more challenging, including demonstration of pharmaceutical equivalence. To understand the underlying science used to evaluate these complex products, we discuss below the standards for review and approval of synthetic generic calcitonin nasal spray products.

Salmon calcitonin, a 32 amino-acid peptide, is used for the effective treatment of osteoporosis. However, one shortcoming regarding the therapeutic use of this peptide drug is its potential to induce unwanted immune responses. Demonstration of pharmaceutical equivalence for salmon calcitonin is particularly important for two reasons: demonstration of the identical active ingredient from two different manufacturers, and control of product and process-related factors that may affect the product immunogenicity. These two key aspects can be addressed through use of prior scientific knowledge, integrated analytical technologies (i.e., an array of properly designed high-resolution analytical characterization methods, biochemical and/or biological assays), and risk assessment, as summarized below.[22]

The active ingredient sameness can be defined by the primary structure or amino acid sequence and a disulfide bond between cysteine residues at positions one and seven, using orthogonal methods.[23,24] Since the immunogenicity of salmon calcitonin is inherent to the active ingredient,[25-27] showing that the generic product has the same salmon calcitonin peptide as the RLD will, at a minimum, ensure that the generic and innovator salmon calcitonin have similar immunogenicity risk. However, there are other product- and process-related factors, such as peptide-related impurities, aggregates, excipients and leachates from a container/closure system that can potentially affect the immunogenicity of a salmon calcitonin nasal spray product.[28,29]

Salmon calcitonin generally has a low tendency for aggregation, and the aggregates are primarily dimers,[30] which are known to be immunogenic. In addition, it has been shown to be pure with low levels of peptide-related impurities. Nevertheless, it is important to characterize aggregate/peptide-related impurities of the salmon calcitonin product, after long-term storage and/or accelerated conditions using orthogonal techniques[31-34] and to show that the generic product has low levels of aggregate/peptide-related impurities, or profiles of aggregate/peptide-related impurities comparable to the RLD.

With regard to formulation, if the generic formulation is qualitatively and quantitatively the same (i.e., using the same excipients and the same concentrations) as the nasal spray formulation of the RLD, then no further study is needed to characterize the excipient effects on the product safety and efficacy because such effects have been evaluated for the RLD. If a different excipient is used in a generic product, it is important to conduct comparative stability studies under the accelerated conditions, and show the comparability of stability against physical and chemical degradations (e.g., aggregates and peptide-related impurities). It is also important to ensure that any difference in excipients does not lead to any significant difference in the excipient–peptide interactions,[35-39] which can affect the conformation and activity of salmon calcitonin.

To mitigate immunogenicity concerns due to leachables, the generic product should use a container/closure system that is compatible with the RDL formulation and minimizes the amount of leachable material throughout shelf life. It is also important to show the absence of significant levels of leachables, or comparability of leachable profiles, between the generic and innovator products using appropriate analytical methods.[40,41]

The above example illustrates the importance of prior knowledge, integrated analytical technologies, and risk assessment in the evaluation of complex drugs, particularly with respect to demonstration of pharmaceutical equivalence. Once bioequivalence is established, the pharmaceutical equivalent is expected to have the same clinical effect and safety profile as the RLD when administered to patients under the condition specified in the labeling. These two drug products may be substituted for each other without any dose adjustment and other additional warning. No additional clinical study is needed.

It should also be noted that this scientific approach can be applicable for evaluation of other complex products, including the so-called NBCDs. Although in some countries a regulatory pathway for the NBCDs is not well defined, in the United States the FDA has defined a regulatory pathway for these drugs, which is 505(j). The FDA has already approved ANDAs for generic drug products, such as enoxaparin injection (2010) and sodium ferric gluconate complex injection (2011). Since their approvals, millions of prescriptions have been filled and no clinical concerns have been observed. In addition, the FDA has published draft guidance for generic doxorubicin HCl liposome injection.

The patient perspective for safety considerations for complex drugs

Janet S. Wyatt (Institute of Pediatric Nursing, and the Arthritis Foundation) described considerations for the safety and efficacy of follow-on versions of complex products from a patient perspective. She noted that the introduction of complex treatments, including biologics and (chemically synthesized) NBCDs, has given a significant benefit to patients with debilitating autoimmune diseases such as rheumatoid arthritis (RA), multiple sclerosis, and psoriasis. Complex drugs, and their follow-on counterparts, are differentiated from small-molecule therapies by their highly complicated structures that may, in some cases, result in a more targeted treatment for some patients. Unfortunately, because of the very same complexity that provides a benefit, some of these drugs can be associated with unforeseen adverse reactions and immunogenic effects, limited availability, and prohibitive costs. To deal with some of these issues, follow-on versions of the drugs need to be properly regulated to ensure that patient safety remains paramount.

Focusing on RA during the New York Academy of Sciences meeting, Wyatt remarked that since the late 1990s, complex drugs of biological origin have been used as a primary form of treatment for 1.3 million patients with RA, a chronic autoimmune disease associated with painful inflammation of the joints and surrounding tissues, leading to bone erosion, deformity, and loss of function.[42] Treatment for RA includes a combination of multiple drugs, including biologics as well as other disease-modifying antirheumatic drugs, analgesics and, often, complementary alternative medicines. While these treatments have helped improve the quality of life for patients, there are a number of treatment inadequacies that still need to be addressed. There is limited knowledge of the impact that gender, race, age, co-morbidities, hormones, environment, and interaction with other drugs will have on patients' responses to treatments. Many patients may experience an unpredictable disease course, and individual responses to treatment may vary. Furthermore, there are no biomarkers in RA to monitor treatment response. As complex biologic products are introduced, therapeutic monitoring for disease and symptom improvement is often complicated, as drug and biologic-related immunogenic factors

need to be taken into account. All of these considerations for patients, which extend to most autoimmune diseases, make the availability of follow-on versions of complex therapies both a promise and a concern.

As patients seek ways to manage and halt disease progression, interest in the development of new and follow-on therapies will grow. However, as new products are developed, it is essential that follow-on versions of complex drugs be proven to work as well as the reference drug. As regulatory authorities begin to embark on the establishment of policy requirements for follow-on versions of complex drugs, these considerations should ensure that patient safety remains a priority. Patients and healthcare providers should view each follow-on product, not as a substitute, but as if the follow-on drug were an entirely different medication with potentially different effects and safety profiles. Clinical studies, including direct comparison trials, may be needed to ensure that the safety and efficacy of the follow-on complex drug is consistent with the reference product. Follow-on medications should be subject to similar regulatory monitoring requirements and post-market surveillance. Finally, patients who receive positive health benefits from their current therapy should not be automatically switched to a follow-on product. As appropriate, all treatment decisions should be established as a result of targeted communication between patients and their healthcare providers. The treatment choices determined by patients and providers should be honored.

As innovations in drug development continue at a rapid pace, efforts are under way to engage all parties in the need to monitor the safety and effectiveness of drugs and complex therapies. Recently the FDA launched its SENTINEL Initiative, with the goal of expanding surveillance of electronic data from healthcare information holders. As patients and caregivers seek timely information about the associated benefits and potential risks of the products they use, increased monitoring of FDA-regulated products will enhance public health safety. As implementation of this initiative is under way, it is hoped that the FDA will seek opportunities to engage patients in programs to improve awareness and education surrounding medications. There is no question that complex drugs have transformed the level of care for patients with autoimmune and other diseases. Looking ahead, it is essential that regulatory authorities ensure a process that will allow follow-on complex drugs to be approved with treatment efficacy and patient safety set as the highest priority. Patients should be educated on the monitoring of the short- and long-term effects of the drugs they are taking, and clear definitions, including instructions on product management and patient monitoring, must be provided to distinguish reference drugs from follow-on versions. With these considerations in mind, patients will continue to benefit from these complex therapies in a safe way that may improve quality of life.

Closing remarks on considerations for NBCDs

To close out the conference, J. Michael Nicholas (Teva Pharmaceuticals) remarked on the importance of the FDA's review of complex drugs, such as NBCDs, on a case-by-case basis, as some generic versions of complex products have already been approved under a 505 (j) regulatory pathway, so that each application is thoroughly evaluated for safety and efficacy. He further stated that in order to ensure that these drugs are safe for patient consumption, the FDA should also consider performing clinical trials when appropriate in order to ensure that adverse effects, for example, the consequences of immunogenicity, do not occur in patients once the drug reaches the market.

Conclusions

The scientific discussions in the conference on NBCDs clearly reflected different opinions regarding these drugs, especially with regard to the regulatory and scientific approaches for reviewing and approving NBCDs. Some participants believed that while the recent biosimilar guidelines released by the FDA earlier this year mark an important milestone in the development of the necessitated regulatory pathways for complex drugs, additional work needs to be conducted to further elucidate a pathway for NBCDs as they currently fall under the 505(j) pathway. In their view, given the complexity of these medications, regulatory authorities should consider defining a new pathway to bring affordable, but safe, versions of NBCDs to market to improve the quality of life for patients. By contrast, others believed that adequate product and process understanding, as well as advances in analytical technologies, can

eliminate the need for clinical studies for evaluation of generic versions of NBCDs.

Acknowledgments

This meeting report presents research featured at the March 9, 2012 conference "Scientific Considerations for Complex Drugs in Light of Established Regulatory Guidance", which was presented by the New York Academy of Sciences in New York City. The conference was supported by Platinum Sponsor Teva Pharmaceutical Industries LTD.

Conflicts of interest

The authors declare no conflicts of interest.

References

1. Mayeno, A.N., F. Lin, C.S. Foote, *et al.* 1990. Characterization of "peak E," a novel amino acid associated with eosinophilia-myalgia syndrome. *Science* **250:** 1707–1708.
2. Wadhwa, M., A.L. Hjelm Skog, C. Bird, *et al.* 1999. Immunogenicity of granulocyte-macrophage colony, stimulating factor (GM-CSF) products in patients undergoing combination therapy with GM-CSF. *Clin. Cancer Res.* **5:** 1353–1361.
3. Boven, K., S. Stryker, J. Knight, *et al.* 2005. The increased incidence of pure red cell aplasia with an Eprex formulation in uncoated rubber stopper syringes. *Kidney Int.* **67:** 2346–2353.
4. Directive 2004/27/EC of the European Parliament and of the Council of 31 March 2004 amending Directive 2001/83/EC on the Community code relating to medicinal products for human use.
5. The Patient Protection and Affordable Care Act (PPAC Act) of 23 March 2010, also referred to in the context of amendments to the Public Health Service Act (PHS Act) as the Biologics Price Competition and Innovation Act (BPCI Act) of 2009.
6. EMEA/190896/2006: Questions and Answers on Recommendation for Refusal of Marketing Application for Alpheon, 28 June 2006.
7. EMEA/CHMP/70349/2008: Withdrawal Assessment Report for Insulin Human Long Marvel.
8. EMA/287731/2011: Questions and Answers, Withdrawal of the marketing authorisation application for Epostim (epoetin alfa), 14 April 2011.
9. EMEA/334517/2009: CHMP (Withdrawal) Assessment Report for Biferonex, July 2009.
10. EMA/CHMP/SWP/100094/2011: Reflection paper on non-clinical studies for generic nanoparticle iron medinal product applications, 17 March 2011.
11. Clark, C. Are all parenteral iron-sucrose preparations identical or just similar? *Hospital Pharmacy Europe*, **58:** Oct 2011, reporting results from Toblli, J.E., G. Cao, J. Giani, et al. Different effects of European iron sucrose similar preparations and originator iron sucrose on nitrosative stress, apoptosis, oxidative stress, biochemical and inflammatory markers in rats. XLVIII ERA EDTA Congress 2011; SuO028.

12. Rottembourg, J., A. Kadri, E. Leonard, *et al.* 2011. Do two intravenous iron sucrose preparations have the same efficacy? *Nephrol. Dial. Transplant.* **26:** 3262–3267.

13. EMEA/CHMP/79769/2006: Reflection Paper on Nanotechnology-Based Medicinal Products for Human Use, 29 June 2006.

14. EMA/CHMP/BMWP/652000/2010: Guideline on similar biological medicinal products containing interferon beta (DRAFT).

15. EMA 2001/83/EC/2001: Directive of the European Parliament and of the Council of 6 November 2001 on the Community Code Relating to Medicinal Products for Human Use.

16. European Medicines Agency (EMA), Committee for Medicinal Products for Human Use (CHMP). 2012. *Guideline on similar biological medicinal products containing monoclonal antibodies—non-clinical and clinical issues* (403543). Retrieved from website: http://www.ema.europa.eu/docs/en_GB/document_library/Scientific_guideline/2012/06/WC500128686.pdf.

17. Toblli, J.E., G. Cao, L. Oliveri & M. Angerosa. 2012. Comparison of oxidative stress and inflammation induced by different intravenous iron sucrose similar preparations in a rat model, Inflamm. *Allergy-Drug Targets* **11:** 66–78.

18. MacNeil. 2009. *WIREs Nanomed Nanobiotechnol 2009.* **1:** 264–271. John Wiley & Sons, Inc.

19. Reflection paper on the data requirements for intravenous liposomal products developed with reference to an innovator liposomal product. Draft, EMA/CHMP/806058/2009, www.ema.europa.eu/docs.

20. Reflection paper on non-clinical studies for generic nanoparticle iron medicinal product applications. March 2011, EMA/CHMP/SWP/100094/2011, www.ema.europa.eu/docs.

21. Orange book: approved drug products with therapeutic equivalence evaluations. Available at http://www.accessdata.fda.gov/scripts/cder/ob/default.cfm (2009). Accessed 17 June 2012.

22. Lee, S.L., L.X. Yu, B. Cai, *et al.* 2010. Scientific considerations for generic synthetic salmon calcitonin nasal spray products. *AAPS J.* **13:** 14–19.

23. Brown, J.R. & B.S. Hartley. 1966. Location of disulphide bridges by diagonal paper electrophoresis. The disulphide bridges of bovine chymotrypsinogen A. *Biochem J.* **101:** 214–228.

24. Creighton, T.E. 1997. *Proteins: structure and molecular properties.* 2nd ed. W. H. Freeman and Company. New York.

25. Kozono, T., M. Hirata, K. Endo, *et al.* 1992. A chimeric analog of human and salmon calcitonin eliminates antigenicity and reduces gastrointestinal disturbances. *Endocrinol.* **131:** 2885–2890.

26. Singer, F.R. 1991. Clinical efficacy of salmon calcitonin in Paget's disease of bone. *Calcif. Tissue Int.* **49:** S7–S8.

27. Rojanasathit, S., E. Rosenberg & J.G. Haddad. 1974. Paget's bone disease: response to human calcitonin in patients resistant to salmon calcitonin. *Lancet* **2:** 1412–1415.

28. Schellekens, H. 2005. Factors influencing the immunogenicity of therapeutic proteins. *Nephrol. Dial. Transplant.* **20:** vi3–9.

29. De Groot, A.S. & D.W. Scott. Immunogenicity of protein therapeutics. *Trends Immunol.* **11:** 482–490.

30. Windisch, V., F. Deluccia, L. Duhau, *et al.* 1997. Degradation pathways of salmon calcitonin in aqueous solution. *J. Pharm. Sci.* **86:** 359–364.

31. Bocian, W., J. Sitkowski, A. Tarnowska, *et al.* 2008. Direct insight into insulin aggregation by 2D NMR complemented by PFGSE NMR. Proteins. **71:** 1057–1065.

32. Kamberi, M., P. Chung, R. DeVas, *et al.* 2004. Analysis of non-covalent aggregation of synthetic hPTH (1–34) by size exclusion chromatography and the importance of suppression of non-specific interactions for a precise quantitation. *J. Chromatogr. B Technol. Biomed. Life Sci.* **810:** 151–155.

33. Lebowitz, J., M.S. Lewis & P. Schuck. 2002. Modem analytical ultracentrifugation in protein science: tutorial review. *Protein Sci.* **11:** 2067–2079.

34. Levin, S. 1991. Field flow fractionation in biomedical analysis. *Biomed. Chromatogr.* **5:** 133–137.

35. Pattnaik, P. 2005. Surface plasmon resonance: applications in understanding receptor-ligand interaction. *Appl. Biochem. Biotechnol.* **126:** 79–92.

36. Bruylants, G., J. Wouters & C. Michaux. 2005. Differential scanning calorimetry in life science: thermodynamics, stability, molecular recognition and application in drug design. *Curr. Med. Chem.* **12:** 2011–2020.

37. Hudson, F.M. & N.H. Andersen. 2004. Exenatide: NMR/CD evaluation of the medium dependence of conformation and aggregation. *State. Biopolymers* **76:** 298–308.

38. Chen, W.L., W.T. Liu, M.C. Yang, *et al.* 2006. A novel conformation-dependent monoclonal antibody specific to the native structure of beta-lactoglobulin and its application. *J. Dairy Sci.* **89:** 912–921.

39. Costantino, H.R., H. Culley, L. Chen, *et al.* 2009. Development of calcitonin salmon nasal spray: similarity of peptide formulated in chlorobutanol compared to benzalkonium chloride as preservative. *J. Pharm. Sci.* **98:** 3691–3706.

40. Ball, D., J. Blanchard, D. Jacobson-Kram, *et al.* 2007. Development of safety qualification thresholds and their use in orally inhaled and nasal drug product evaluation. *Toxicol. Sci.* **97:** 226–236.

41. Norwood, D.L., D. Paskiet, M. Ruberto, *et al.* 2008. Best practices for extractables and leachables in orally inhaled and nasal drug products: an overview of the PQRI recommendations. *Pharm. Res.* **25:** 727–739.

42. Helmick, C.G. *et al.* 2008. Estimates of the prevalence of arthritis and other rheumatic conditions in the United States. Part I. *Arthritis Rheum.* **58:** 15–25. doi: 10.1002/art.23177.

Appendix

Agenda of the conference

8:00 am

Breakfast and registration

9:00 am

Opening remarks

Brooke Grindlinger, PhD, The New York Academy of Sciences

9:15 am
  Keynote Address
  Small Molecules, Biologics and Non-biological Complex Drugs: An Overview of their Scientific and Regulatory Differences
  Chris Holloway, PhD, ERA Consulting Group
10:00 am
  Biosimilar Monoclonal Antibodies: Current European Developments
  Jan Mueller-Berghaus, MD, Paul-Ehrlich-Institut
10:30 am
  Networking Coffee Break
11:00 am
  Insights Into the Process of Establishing Regulations for NBCDs
  Beatriz Silva-Lima, PhD, University of Lisbon
11:30 am
  Experimental and Clinical Trial Evidence Needed for Follow-On Versions of Complex Drugs
  Huub Schellekens, MD, PhD, Utrecht University
12:00 pm
  Therapeutic Equivalence of Complex Drug Products: Case Study of Generic Calcitonin Nasal Spray Products
  Sau (Larry) lee, PhD, U.S. Food and Drug Administration
12:30 pm
  Scientific Challenges in Regulating Complex Biological Products in Emerging Markets
  Ivana knezevic, MD, PhD, World Health Organization
1:00 pm
  Networking lunch – Speaker-guided Discussion groups Topic: Defining Complex Drugs
2:30 pm
  Panel Discussion – *Defining Complex Drugs*
  Moderator: Daan J. A. Crommelin, PhD, Utrecht University
3:15 pm
  Clinical Barriers to the Adoption of Biosimilars: A Lesson for Complex Drugs
  Andrew D. Zelenetz, MD, PhD, Memorial Sloan-Kettering Cancer Center
3:45 pm
  Follow-on Biologics – A Patient's Perspective
  Janet S. Wyatt, PhD, RN, FAANP, Institute of Pediatric Nursing and Arthritis Foundation
4:15 pm
  Networking Coffee Break
4:45 pm
  Panel Discussion – *Physician and Patient Opinion*
5:15 pm
  Closing remarks
  J. Michael Nicholas, PhD, Teva Pharmaceuticals
5:30 pm
  Adjourn

Ann. N.Y. Acad. Sci. ISSN 0077-8923

ANNALS OF THE NEW YORK ACADEMY OF SCIENCES
Issue: Annals *Meeting Reports*

Fetal programming and environmental exposures: implications for prenatal care and preterm birth

Thaddeus T. Schug,[1] Adrian Erlebacher,[2] Sarah Leibowitz,[3] Liang Ma,[4] Louis J. Muglia,[5] Oliver J. Rando,[6] John M. Rogers,[7] Roberto Romero,[8] Frederick S. vom Saal,[9] and David L. Wise[10]

[1]National Institute of Environmental Health Sciences, Research Triangle Park, North Carolina. [2]NYU School of Medicine, New York, New York. [3]The Rockefeller University, New York, New York. [4]Division of Dermatology, Department of Medicine, Washington University School of Medicine, St. Louis, Missouri. [5]Cincinnati Children's Hospital Medical Center and University of Cincinnati College of Medicine, Cincinnati, Ohio. [6]Department of Biochemistry and Molecular Pharmacology, University of Massachusetts Medical School, Worcester, Massachusetts. [7]Toxicity Assessment Division, National Health and Environmental Effects Research Laboratory, Office of Research and Development, United States Environmental Protection Agency, Research Triangle Park, North Carolina. [8]Perinatology Research Branch, NICHD, NIH, DHHS, Bethesda, Maryland. [9]Division of Biological Sciences, University of Missouri, Columbia, Missouri. [10]Merck Research Laboratories, West Point, Pennsylvania

Address for correspondence: Thaddeus T. Schug, National Institute of Environmental Health Sciences, P.O. Box 12233, Research Triangle Park, NC 27709. schugt@niehs.nih.gov

Sponsored by the New York Academy of Sciences and Cincinnati Children's Hospital Medical Center, with support from the National Institute of Environmental Health Sciences (NIEHS), the National Institute on Drug Abuse (NIDA), and Life Technologies, "Fetal Programming and Environmental Exposures: Implications for Prenatal Care and Preterm Birth" was held on June 11–12, 2012 at the New York Academy of Sciences in New York City. The meeting, comprising individual talks and panel discussions, highlighted basic, clinical, and translational research approaches, and highlighted the need for specialized testing of drugs, consumer products, and industrial chemicals, with a view to the unique impacts these can have during gestation. Speakers went on to discuss many other factors that affect prenatal development, from genetics to parental diet, revealing the extraordinary sensitivity of the developing fetus.

Keywords: genetic and epigenetic programming; fetal development; toxicity

Background and perspectives

Keynote address: Frederick S. vom Saal
(University of Missouri–Columbia)

Fetal programming is an enormously complex process that relies on numerous environmental inputs from uterine tissue, the placenta, the maternal blood supply, and other sources. Recent evidence has made clear that the process is not based entirely on genetics, but rather on a delicate series of interactions between genes and the environment. It is likely that epigenetic ("above the genome") changes are responsible for modifying gene expression in the developing fetus, and these modifications can have long-lasting health impacts. Determining which epigenetic regulators are most vital in embryonic development will improve pregnancy out-comes and our ability to treat and prevent disorders that emerge later in life.

"Fetal Programming and Environmental Exposures: Implications for Prenatal Care and Preterm Birth" began with a keynote address by Frederick vom Saal, who explained that low-level exposure to endocrine disrupting chemicals (EDCs) perturbs hormone systems *in utero* and can have negative effects on fetal development. vom Saal presented data on the EDC bisphenol A (BPA), an estrogen-mimicking compound found in many plastics. He suggested that low-dose exposure to EDCs can alter the development process and enhance chances of acquiring adult diseases, such as breast cancer, diabetes, and even developmental disorders such as attention deficit disorder (ADHD).[1] vom Saal noted that conventional risk-assessment

doi: 10.1111/nyas.12003
Ann. N.Y. Acad. Sci. 1276 (2012) 37–46 © 2012 New York Academy of Sciences.

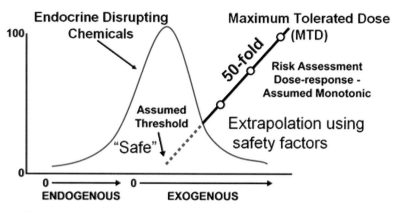

Figure 1. Chemical risk assessment assumes a monotonic dose–response relationship and makes assumptions about low-dose effects.[2]

strategies used for most chemicals cannot be used to assess toxicity of endocrine-active compounds, which often display nonmonotonic dose responses (Fig. 1).[2]

Session I: Genetic/epigenetic programming of preimplantation development

Chair: Marco Conti (University of California, San Francisco)

Marco Conti spoke about networks of RNA-binding proteins that regulate oocyte and embryonic development. Conti focused on the mechanism by which newly synthesized proteins "set the stage" for the developing embryo between the stages of oocyte maturation and activation of the embryonic genome. Conti discussed the discovery of the gene *Dazl* (deleted in azoospermia-like), which encodes a maternal protein that positively regulates RNA translation in the maturing oocyte before embryogenesis occurs.[3] These effects are due to epigenetic effects on DNA and chromatin structure. Conti's research showed that the absence of DAZl prevents meiosis by disrupting microtubule organization and spindle formation.

Kelle H. Moley (Washington University School of Medicine) presented her work in a mouse model of type 1 diabetes, in which oocytes are smaller and show impaired mitochondrial function and aberrant morphology, as well as abnormal meiotic spindle formation and aneuploidy.[4] Transfer of these fertilized oocytes into nondiabetic mice

results in poor reproductive outcomes, including growth restriction and congenital anomalies,[5] suggesting that mitochondrial dysfunction leads to either aneuploidy and embryo arrest or fetal growth abnormalities in the next generation.

Similarly in mouse models of obesity, exposure to an abnormal endocrine environment affects the oocyte, the embryo, and pregnancy outcomes. Moley noted that oocytes from mice maintained on a high-fat diet are significantly smaller, show delayed meiotic maturation, and show impaired mitochondrial function and aberrant morphology, as well as abnormal meiotic spindle formation and aneuploidy. Transfer of these fertilized oocytes into nondiabetic mice results in poor reproductive outcomes, including growth restriction and brain anomalies.[6] These findings suggest that metabolic changes in the mother can affect development as early as the unfertilized oocyte, and that these changes may have long-term effects on offspring.[6] In humans, as well, abnormal levels of free fatty acids in follicular fluid, as well as sera, have been associated with poor oocyte quality and decreased chance of pregnancy in patients undergoing in vitro fertilization (IVF), respectively.[7,8]

The session concluded with a presentation from James Charles Cross (University of Calgary). Cross examined cells of the placenta–fetus interface and found that several types of trophoblast cells line the canal leading to the umbilical cord. The fetus relies on these cells for transportation of maternal blood and nutrients. Specialized placental cells express a family of genes that encode for prolactins, hormones required for insulin production during pregnancy.[9] Cross noted that prolactin hormones protect female mice from developing gestational diabetes during pregnancy, further illustrating the importance of the specialized cells that bridge the gap between mother and fetus.

Session II: Embryo–uterine crosstalk

Chair: Adrian Erlebacher (New York University Langone Medical Center)

The second session began with a presentation by Francesco DeMayo (Baylor College of Medicine). DeMayo spoke on the role progesterone receptors play in embryo implantation and decidualization. DeMayo noted that complex signaling events occur between the epithelium and the stroma before implantation.[10] His group discovered that activation of the progesterone receptor in the epithelium sets off this cascade by regulating expression of the morphogen Indian Hedghog (Ihh). His group is also seeking to understand the mechanism by which progesterone transforms the endometrium during pregnancy.

DeMayo also spoke about the effects of the epidermal growth factor receptor (EGFR) family of receptor tyrosine kinases on decidualization.[11] His group found that progesterone action is regulated by EGFR family members. DeMayo explained that pharmacological inhibitors of EGFR and HER2 prevent decidualization and may be used as a therapeutic to treat endometriosis. DeMayo's group is undertaking genome-wide studies to identify which genes are targets of the progesterone receptor and to further define how uterine biology is dependent on this hormone.[12] Preliminary findings indicate the importance of genes involved in the circadian cycle and of the GATA2 gene, which encodes a transcription factor that tunes progesterone sensitivity in the epithelium.

Bruce Murphy (Université de Montréal) spoke next about a family of genes called orphan nuclear receptors, which are similar to estrogen and progesterone receptors, except that no activating hormones have yet been identified (hence the term *orphan receptor*). The *NR5A2* gene encodes an orphan receptor that is expressed in granulosa and luteal cells in the ovary and the uterine stroma.[13] Murphy and colleagues are investigating this orphan receptor using tissue-specific knockout mice with the *NR5A2* gene

deleted in progesterone-receptive tissues only. His group discovered that *NR5A2* deletion causes problems early in gestation by inhibiting progesterone synthesis. Murphy noted that knockout mice treated with progesterone exhibited gestational failure later, due to defects in the uterus. Gene-expression experiments suggest that the *NR5A2* gene may play an important role in decidualization by regulating networks of signaling molecules and transcription factors that are required for fertility. While much attention has been directed toward understanding the action of well-established hormone receptors, these results indicate the importance of orphan nuclear receptors in regulating reproductive processes.

Adrian Erlebacher (NYU Langone Medical Center) ended the session venturing into the field of immunology. Erlebacher began by explaining why the immune system fails to reject a fetus. To investigate this, Erlebacher reactivated memory T cells in a transgenic mouse model and asked whether these cells could migrate to the maternal–fetal interface. His team found that T cells cannot accumulate within the decidua, suggesting that there is a molecular barrier preventing infiltration; they discovered that an innate epigenetic mechanism represses certain chemokine genes that are required for T cell migration.[14] Erlebacher speculated that insufficient epigenetic silencing of chemokine genes would result in fetal/placental "rejection" or preterm birth secondary to deciduitis. Conversely, excessive silencing would result in an inability to combat decidual infection and lead to preterm birth secondary to chorioamnionitis or to stillbirth.

Session III: Presentations from young investigators

Chair: Sudhansu K. Dey (Cincinnati Children's Hospital)

In the next session, graduate students and postdoctoral fellows presented their research projects in fetal programming in a series of presentations. The first talk was given by Jenny Sones (Cornell University), who presented work on a mouse model called BPH/5, an inbred strain that spontaneously develops symptoms similar to human preeclampsia, including late-gestational hypertension, proteinuria, renal glomerular lesions, and endothelial dysfunction.[15] She discovered that the model mice had inappropriate hormone dynamics, which led to abnormal expression of two important genes that regulate implantation, *LIF* and *COX2*.

Jeeyeon Cha from S. K. Dey's group (Cincinnati Children's Hospital Medical Center) presented work that showed induction of preterm birth in mice conditionally deleted of uterine Tr p53 than encodes p53. Cha and colleagues showed that the important enzymatic complex mTORC1 is upregulated in decidual cells and induces their senescence—premature exit from the cell cycle, resulting in heightened Cox2 activity in the decidua. Hypothesizing that preterm birth is characterized by senescent decidual cells that have undergone premature terminal differentiation (Fig. 2), they found that preterm birth is reversed by a selective Cox2 inhibitor, celecoxib, or by a low dose of the anti-inflammatory and anti-immunosuppressive drug rapamycin, which is commonly used to inhibit mTORC1 activity. This work is clinically relevant and could one day facilitate novel preventative treatment strategies for women at risk for preterm birth.

Alicia Bárcena (University of California, San Francisco) presented her work on hematopoietic cells in the placenta and the chorion.[16] She discovered that a certain type of hematopoietic stem cell exhibiting high expression of a primitive cell surface marker is capable of repopulating the extraembryonic niche with diverse blood cells; these stem cells also communicate with endothelial cells in the vasculature, as well as with trophoblasts in the chorion. This work will further our understanding of how primitive blood development occurs early in embryogenesis.

Finally, Martha Susiarjo (University of Pennsylvania) presented her work on the effects of BPA in the mouse placenta. Susiarjo's talk revisited the role of BPA in altering epigenetic regulation and causing developmental defects and other diseases. She hypothesized that BPA alters DNA methylation of imprinted genes, leading to aberrant expression. She found that low-dose exposure leads to loss of imprinting of several genes at multiple domains, including *Snrpn* and *Kcnq1*. These gene-expression changes were associated with DNA hypomethylation, an epigenetic mechanism linked to gene silencing.[17,18] Future studies will determine the molecular mechanisms involved in BPA-induced DNA methylation changes, as well as in the developmental effects of exposure in the placenta.

Session IV: Decidualization and placentation

Chair: R. Michael Roberts (University of Missouri–Columbia)

Susan J. Fisher (the University of California, San Francisco) began the next session describing how trophoblast development and invasion are essential steps in placentation, a process that establishes a fetal connection to maternal blood flow. Fisher noted that species differences in this developmental process have highlighted the need to establish better model systems to study human trophoblast invasion. Fisher described a new method to derive human embryonic stem cell (hESC) lines from single blastomeres from eight-cell embryos rather than from whole blastocysts.[19] The human embryonic stem cell is a powerful model for understanding human development, but lines that are currently available have been established after weeks of *in vitro* development and thus exhibit heterogeneity

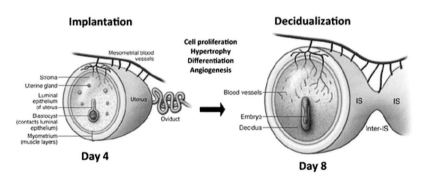

Figure 2. Cellular senescence and premature differentiation of decidual cells may lead to defective decidualization and preterm birth. The above scheme is adapted from Ref. 46.

with reference to gene expression and differentiation capacity. Fisher's new hESC lines appear to be more primitive and display a slightly different gene-expression profile, with fewer genes expressed than previously established hESC lines (these new cell lines are awaiting National Institutes of Health (NIH) registration). Experiments are underway to differentiate these lines into trophoblast cells for studying human trophoblast development.

Michael J. Soares (Kansas University Medical Center) described several models for studying trophoblast invasion. Soares investigated the molecular pathways that regulate trophoblast invasion using both transgenic rats, a model system with an invasive trophoblast lineage similar to that of humans, and blastocyst-derived trophoblast stem cell lines.[20] Fosl1 and hypoxia inducible factor (HIF)–signaling pathways regulate trophoblast invasion in rats.[21,22] Fosl1 encodes a transcription factor that is a component of the PI3K/Akt signaling pathway, which is important in promoting cell growth and survival and in trophoblast cells to regulate invasiveness and the expression of matrix metalloproteinase 9 (MMP9).[22] Activated HIF induces the expression of several genes, including the histone demethylase Kdm3A, which regulates trophoblast invasion and the expression of another matrix metalloprotease, MMP12.[21] Members of the MMP family are key regulators of cell adhesion, extracellular matrix remodeling, and invasion.

R. Michael Roberts (University of Missouri-Columbia) continued the discussion of trophoblast invasion and development. His results differed from those reported showing that hESCs treated with the growth factor BMP4 differentiate along the mesodermal lineage.[23] Roberts' data show, instead, that treating hESCs with BMP4 under standard culture conditions switches on expression of trophoblast markers and minimal upregulation of mesoderm markers. The Matrigel[TM] invasion assay revealed that cells became increasingly invasive over time, especially in an environment high in oxygen (20%). Roberts suggested that the different results could be explained by the use of different media and substrata by the two groups to drive differentiation. His group has now derived 14 human-induced pluripotent stem cell (hiPSC) lines from fetal umbilical cord tissue (obtained from cases of preeclampsia, a condition that can develop during pregnancy and can result in the death of the baby and/or mother from seizures and organ failure). Roberts' team plans to use the hiPSC lines to investigate possible genetic and cellular defects related to preeclampsia.

Session V: Environmental contributors: drugs, diet, and consumer products

Chair: Margaret Ann Miller (FDA/National Center for Toxicological Research)
Keynote Speaker: Sarah Leibowitz (The Rockefeller University)

The second day of the conference highlighted the impact of diet and drugs on fetal programming and presented current research into the causes and prevention of preterm birth. It is clear that parental diet can affect fetal development: the challenge is to understand its effects and to pinpoint periods of gestation during which it is most influential on health outcomes. John R. G. Challis (University of Toronto) discussed a prenatal study of individuals whom experienced the Dutch winter famine.[24] The study found, compared to baseline, that prenatal malnutrition caused numerous changes that influence disease, including decreased glucose tolerance and elevated insulin concentrations. Further analyses revealed epigenetic alterations including decreased DNA methylation of the *IGF2* gene. These findings highlight the significant effect that maternal nutrition during gestation has on health later in life.[25]

Data showing that paternal diet also affects offspring metabolism were presented by Oliver J. Rando (University of Massachusetts Medical School). His group is developing a mouse epigenetic pipeline to determine whether dietary information is carried by sperm.[26] Using sperm from males with varied diets to fertilize eggs *in vitro*, researchers derive embryonic stem cells from developing blastocysts for further *in vitro* studies or to implant these blastocysts into pseudo-pregnant females for *in vivo* studies. This strategy eliminates other factors that could potentially affect offspring metabolism, such as seminal fluid peptides, and, in theory, enables researchers to study only the epigenetic impact of the sperm in an otherwise genetically homogenous population. Thus far, analysis of the genetic and epigenetic landscape of sperm has revealed modest cytosine methylation changes at putative gene enhancer regions, which could influence gene expression. Although methylation patterns generally clustered better with the generation of mice than with their diet, the interferon zeta cluster displayed a 10 to 15% decrease in methylation with a low protein diet. Research is ongoing to discover the implications of these results and to further refine the technique used.

Pat Levitt (University of Southern California) continued the discussion of metabolic pathways and fetal development, focusing on the role played by the placenta. There is a great deal of evidence supporting the hypothesis that maternal serotonin (5HT) affects fetal brain development by directly influencing serotonin levels in the developing brain.[27] Levitt showed that 5HT produced by the placenta contributes to serotonin levels in the developing brain: in Pet-1 knockout mice 5HT levels in the forebrain remained normal until embryonic day (E) 16.5, despite the lack of 5HT neurons. Demonstrating that the placenta is capable of producing 5HT until E18.5, Levitt's data shed light on the importance of this extraembryonic source of a neurotransmitter for fetal development.

John Schimenti (Cornell University) presented work using forward and reverse genetic approaches to identify genes and pathways that influence germ cell survival.[28] Germline mutations, if not corrected or eliminated by the DNA damage response (DDR) pathway, can result in birth defects, reduced fitness, and sterility. Intact cell-death machinery facilitates the elimination of germ cells that fail to be corrected by the DDR. Targeted deletion of either *Mcm9* or *Fancm*, two DDR pathway genes that are needed for correcting mutations, resulted in germ-cell depletion in the ovaries and testes of mice.[29] Deletion of the p53 gene, which triggers cell death in response to DNA damage, resulted in germ-cell depletion in *Mcm9*-knockout mice but not in *Fancm*-knockout mice. Analysis of a number of other knockouts revealed that germ cell depletion in response to DNA damage can occur via p53-dependent, as well as p53-independent and Chek2-independent mechanisms. This provides evidence that several molecular pathways are at work to cull germ cells carrying mutations that could result in embryos at risk for birth defects and sterility.

Sarah Leibowitz focused on the positive feedback loop between maternal diet during gestation and the dietary habits of offspring over their lifetime.

In one study in rats, a high-fat diet and high alcohol consumption were both correlated with increased production of hypothalamic peptides in offspring, which drive an appetite for a high-fat diet. Prenatal exposure to a high-fat diet increased the genesis of peptide-expressing neurons in the hypothalamus of developing rats at embryonic day (E) 11.5, and offspring exhibited a higher density of these neurons after birth.[30] This higher density persisted even in the absence of a high-fat diet over their lifetime and resulted in a range of behavioral and metabolic effects, such as a higher food intake and a preference for fatty foods. This study underscores the significant impact of maternal diet during pregnancy on the dietary habits of offspring.

Session VI: Pregnancy disorders and prematurity: gene–environment interactions

Chair: S. Ananth Karumanchi (Beth Israel Deaconess Medical Center and Harvard Medical School)

Louis J. Muglia (Cincinnati Children's Hospital Medical Center) began the next session discussing factors influencing preterm delivery. Inactivation of cyclooxygenase 1 (Cox1)—a key regulator of prostaglandin biosynthesis—is implicated in human parturition-delayed birth in mice; however, implantation of Cox1-deficient blastocysts into the uterus of normal females resulted in normal duration of labor, suggesting that maternal Cox1 expressed in the uterine wall is both necessary and sufficient for normal labor.[31] Muglia uses comparative and whole-genome studies to identify gene variants in humans that may contribute to preterm labor;[32,33] this could allow for prenatal identification of at-risk individuals and help researchers to develop new therapies for treatment.

Roberto Romero (Eunice Kennedy Shriver National Institute of Child Health and Human Development /NIH) reviewed the concept of preterm parturition as a syndrome and the importance of progesterone for pregnancy maintenance. The administration of progesterone receptor antagonists induces cervical ripening in animals and in pregnant women in the first, second, and third trimesters. He described the relationship between a sonographic short cervix, detected by ultrasound in the midtrimester of pregnancy, and the risk for spontaneous preterm delivery. For example, an asymptomatic patient with a cervix of 15 mm or less has a 50% risk of preterm delivery of <32 weeks. Two randomized clinical trials in which vaginal progesterone was used to prevent preterm delivery in women with a short cervix were presented. The results of the PREGNANT Trial (international trial)[34] showed that the administration of 90 mg of progesterone/day (vaginal route) reduced the rate of preterm delivery of <33 weeks by 45% and the rate of respiratory distress syndrome by 61%. The results of an individual patient meta-analysis [35] were presented in which vaginal progesterone administration reduced the rate of preterm delivery at <28, <33, and <35 weeks of gestation. This approach was also associated with a reduction in the number of admissions to newborn intensive care units, respiratory distress syndrome, the need for mechanical ventilation, and a composite score of neonatal morbidity and mortality. It is estimated that universal screening to identify a short cervix and treatment of this condition with vaginal progesterone would save $19 million per 100,000 women screened, and the total estimate of savings of the U.S. healthcare system would be $500–$700 million per year.[36] This effect is not achieved by using 17-α-hydroxyprogesterone caproate, which is a synthetic progestin with a potential safety signal.

S. Ananth Karumanchi (Beth Israel Deaconess Medical Center and Harvard Medical School) talked about the role of several angiogenic growth factors and their inhibitors that contribute to preeclampsia in women who present prematurely with severe disease.[37] In particular, he focused his attention on vascular endothelial growth factor (VEGF) signaling and discussed how impairment of VEGF signaling may lead to preeclamptic signs and symptoms. He also discussed strategies to treat preeclampsia by depleting proteins that interfere with critical angiogenic factors, such as VEGF, to shift the balance of factors toward proangiogenesis.

Session VII: Environment, hormone action, and pharmacological/endocrine disruptors

Chair: L. David Wise (Merck)

There are numerous consumer products on the market for which safety data are lacking. The FDA, academic institutions, and nongovernmental organizations are conducting studies to address the safety of these compounds for fetal development and postnatal health. Liang Ma (Washington University School of Medicine) described his research into the mechanism of action of diethylstilbestrol (DES), a synthetic estrogen that was widely used as an anti-miscarriage drug between 1940 and 1970 but was later shown to induce reproductive tract malformations and other defects.[38] Ma's data show that DES represses key developmental genes such as *Hoxa10*, inhibits cell death, reduces cell proliferation in the uterine epithelium, induces accumulation of lipid droplets, and promotes upregulation of the glucose pathway. Ma predicts that the metabolic defects caused by DES exposure could also be present in other estrogen-responsive tissues, such as fatty tissue, and thus may contribute to obesity later in life.[39]

Patricia Hunt (Washington State University) discussed reproductive tract abnormalities induced by BPA. Hunt's investigation of BPA was prompted by an observation that chromosomal abnormalities occur in mice that consume water from BPA bottles.[40] Treating mice with BPA at fetal and perinatal stages increased chromosomal recombination, increased multioocyte follicle formation in females, and affected spermatogenesis in males. Similar effects were detected in experiments in rhesus monkeys.[41] Future studies will analyze the effect of BPA on other organs, including the brain.

L. David Wise described the guidelines issued by the International Conference on Harmonization (http://www.ich.org/products/guidelines/safety/article/safety-guidelines.html) that outline a series of studies set by the United States, the European Union, and Japan for examining the effects of drugs on several stages of prenatal and postnatal development. These guidelines cover studies on fertility and early development (from fertilization until implantation), embryo–fetal development (from implantation until just before birth), and prenatal and postnatal development (from implantation until weaning). Additional guidelines address the need and design of juvenile toxicity (from birth until after sexual maturity) studies. Wise's talk detailed the design and types of data collected in the embryo–fetal studies in rodents and rabbits.[42] Although labor intensive, these studies provide the most suitable data for determining risk information for women of childbearing age.

Sessions VIII and IX: Panel and general audience discussion

Moderator: Patricia Hunt (Washington State University)

Energy remained high during the meeting's final session, with panel members and the audience deliberating on the question, How can we better predict, assess, and lower the risk of, and/or prevent environmental and genetic factors that predispose to, adverse fetal outcomes and preterm birth? John Rogers (U.S. EPA) pointed out the challenges of developing defined *in vitro* and *in vivo* tests to better assess developmental exposures over critical windows of development. Rogers noted that it is also important to agree upon what defines adverse endpoints, such as gene expression, phenotypic changes, and functional developmental changes. He also noted that it is important to incorporate functional challenges to reveal effects, for example, providing a high-fat diet for studies of putative environmental obesogens. Fred vom Saal went on to add that we need to incorporate appropriate developmental assays that include long-latency outcomes.

Thad Schug (National Institute of Environmental Health Sciences (NIEHS)) explained how a collaborative group of environmental health scientists and green chemists have developed a tiered endocrine disruption screening protocol (TiPED) (Fig. 3) that is very different from current regulatory endocrine disruption screening programs.[43] The protocol is designed for chemists to use in early stages of chemical development and is completely voluntary. It involves a time- and cost-sensitive approach to toxicology testing, with assay complexity and cost increasing as one proceeds through testing tiers. The protocol starts with computer modeling and ends with mammalian testing. A positive test anywhere in the protocol means the chemical is problematic and needs further evaluation or redesign. The group has developed guiding principles for assays and guidelines for proper laboratory testing. The protocol will be online on December 6, 2012 and in print in January 2013 in the U.K. Royal Society of Chemistry's *Green Chemistry* journal; a real-time online version will soon be available.

Schug noted that the NIEHS, the National Toxicology Program (NTP), and the FDA have formed the Consortium Linking Academic and Regulatory Insights on BPA Toxicity (CLARITY-BPA) program to support a perinatal two-year good laboratory practices (GLP) chronic toxicity study on BPA.[44] In addition to the core elements of a GLP-compliant study, the study involves academic research partners and will incorporate a wide range of doses and disease-relevant endpoints that have not been used in any previous GLP-compliant BPA toxicity study. The consortium is making all experimental data available via the Chemical Effects in Biological Systems (CEBS) database system developed by the National Toxicology Program at NIEHS. CEBS is a public resource, comprising integrated depositions of data from academic, industry, and governmental laboratories,[45] and a tool designed to better coordinate multilevel studies and support meta-analysis.

Toward the end of the session members of the panel and the audience engaged in an active discussion on the experimental challenges for determining the health implications of low-dose effects and nonmonotonic dose responses in relation to endocrine-active chemicals.[2] Debate centered on whether current observations in the literature are sufficient to reexamine the ways in which chemicals are tested for endocrine-disrupting properties and how risk to human health may be managed. To this end, the NEIHS cosponsored a workshop[a] that

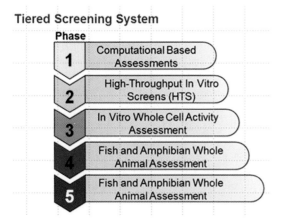

Tiered Screening System

Phase

1	Computational Based Assessments
2	High-Throughput In Vitro Screens (HTS)
3	In Vitro Whole Cell Activity Assessment
4	Fish and Amphibian Whole Animal Assessment
5	Fish and Amphibian Whole Animal Assessment

Figure 3. Tiered protocol for endocrine disruption (TiPED). The approach to using this tiered system runs from the simplest, fastest, and cheapest on the top (Tier 1) to the most expensive on the bottom (Tier 5). Failure to find EDC activity in one tier then leads to testing at the next level tier.

---

[a]http://www.niehs.nih.gov/about/visiting/events/pastmtg/2012/dert_endocrine/index.cfm

aimed to define research needs required to move closer to scientific agreement on low-dose effects and nonmonotonic dose responses for endocrine-active substances. Rogers noted that the EPA has created a workgroup to develop a response to a recent article on low-dose nonmonotonic dose responses, and an agency position on how such dose responses should be considered for risk-assessment purposes.

In her closing remarks, Fisher said the research presented at the meeting will ultimately lead to developments that "will improve pregnancy outcomes and our ability to treat and prevent disorders that emerge later in life." "It's exciting to see how far we have come in our understanding of the complex interactions between genes and our environment occurring *in utero*," added Fisher.

Acknowledgments

This meeting report presents research featured at the June 11–12, 2012 Fetal Programming and Environmental Exposures: Implications for Prenatal Care and Preterm Birth conference, presented by the New York Academy of Sciences and Cincinnati Children's Hospital Medical Center. The conference was supported by Cincinnati Children's Hospital Medical Center and Academy Friend Sponsors, Abcam Inc., Quartzy, Watson Pharmaceuticals, Inc., and the Ronald O. Perelman and Claudia Cohen Center for Reproductive Medicine at Weill Cornell Medical College. The conference was also supported by educational grants from the Burroughs Wellcome Fund and the March of Dimes Foundation (Grant No. 4-FY12–545). Funding for this conference was also made possible, in part, by Grant number 1R13ES021699–01 from the National Institute of Environmental Health Sciences (NIEHS) and the National Institute on Drug Abuse (NIDA). The views expressed in written conference materials or publications and by speakers and moderators do not necessarily reflect the official policies of the Department of Health and Human Services, nor does mention by trade names, commercial practices, or organizations imply endorsement by the U.S. government.

Conflicts of interests

The authors declare no conflicts of interest.

References

1. vom Saal, F.S. *et al.* 2012. The estrogenic endocrine disrupting chemical bisphenol A (BPA) and obesity. *Mol. Cell. Endocrinol.* **354:** 74–84.
2. Vandenberg, L.N. *et al.* 2012. Hormones and endocrine-disrupting chemicals: low-dose effects and nonmonotonic dose responses. *Endocr. Rev.* **33:** 378–455.
3. Chen, J. *et al.* 2011. Genome-wide analysis of translation reveals a critical role for deleted in azoospermia-like (Dazl) at the oocyte-to-zygote transition. *Genes. Dev.* **25:** 755–766.
4. Wang, Q. *et al.* 2009. Maternal diabetes causes mitochondrial dysfunction and meiotic defects in murine oocytes. *Mol. Endocrinol.* **23:** 1603–1612.
5. Wyman, A. *et al.* 2008. One-cell zygote transfer from diabetic to nondiabetic mouse results in congenital malformations and growth retardation in offspring. *Endocrinology* **149:** 466–469.
6. Jungheim, E.S. *et al.* 2010. Diet-induced obesity model: abnormal oocytes and persistent growth abnormalities in the offspring. *Endocrinology* **151:** 4039–4046.
7. Jungheim, E.S. *et al.* 2011. Associations between free fatty acids, cumulus oocyte complex morphology and ovarian function during in vitro fertilization. *Fertil. Steril.* **95:** 1970–1974.
8. Jungheim, E.S. *et al.* 2011. Elevated serum alpha-linolenic acid levels are associated with decreased chance of pregnancy after in vitro fertilization. *Fertil. Steril.* **96:** 880–883.
9. Hu, D. & J.C. Cross. 2010. Development and function of trophoblast giant cells in the rodent placenta. *Int. J. Dev. Biol.* **54:** 341–354.
10. Franco, H.L. *et al.* 2012. Epithelial progesterone receptor exhibits pleiotropic roles in uterine development and function. *FASEB J.* **26:** 1218–1227.
11. Jin, N. *et al.* 2009. Mig-6 is required for appropriate lung development and to ensure normal adult lung homeostasis. *Development* **136:** 3347–3356.
12. Rubel, C.A. *et al.* 2012. GATA2 is expressed at critical times in the mouse uterus during pregnancy. *Gene. Expr. Patterns* **12:** 196–203.
13. Duggavathi, R. *et al.* 2008. Liver receptor homolog 1 is essential for ovulation. *Genes. Dev.* **22:** 1871–1876.
14. Nancy, P. *et al.* 2012. Chemokine gene silencing in decidual stromal cells limits T cell access to the maternal-fetal interface. *Science* **336:** 1317–1321.
15. Hoffmann, D.S. *et al.* 2008. Chronic tempol prevents hypertension, proteinuria, and poor feto-placental outcomes in BPH/5 mouse model of preeclampsia. *Hypertension* **51:** 1058–1065.
16. Barcena, A. *et al.* 2011. Human placenta and chorion: potential additional sources of hematopoietic stem cells for transplantation. *Transfusion* **51**(Suppl 4): 94S–105S.
17. Feinberg, A.P. & B. Tycko. 2004. The history of cancer epigenetics. *Nat. Rev. Cancer* **4:** 143–53.
18. Arnaud, P. & R. Feil. 2005. Epigenetic deregulation of genomic imprinting in human disorders and following assisted reproduction. *Birth Defects Res. C. Embryo. Today* **75:** 81–97.

19. Aghajanova, L. *et al.* 2012. Comparative transcriptome analysis of human trophectoderm and embryonic stem cell-derived trophoblasts reveal key participants in early implantation. *Biol. Reprod.* **86:** 1–21.

20. Soares, M.J. *et al.* 2012. Rat placentation: an experimental model for investigating the hemochorial maternal-fetal interface. *Placenta* **33:** 233–243.

21. Chakraborty, D. *et al.* 2011. Natural killer cells direct hemochorial placentation by regulating hypoxia-inducible factor dependent trophoblast lineage decisions. *Proc. Natl. Acad. Sci. USA* **108:** 16295–16300.

22. Kent, L.N. *et al.* 2011. FOSL1 is integral to establishing the maternal-fetal interface. *Mol Cell Biol.* **31:** 4801–4813.

23. Bernardo, A.S. *et al.* 2011. BRACHYURY and CDX2 mediate BMP-induced differentiation of human and mouse pluripotent stem cells into embryonic and extraembryonic lineages. *Cell Stem Cell* **9:** 144–155.

24. Challis, J. 2001. Glucose tolerance in adults after prenatal exposure to famine. *Lancet* **357:** 1798.

25. Challis, J.R. 2012. Endocrine disorders in pregnancy: Stress responses in children after maternal glucocorticoids. *Nat. Rev. Endocrinol.* **8:** 629–630.

26. Carone, B.R. *et al.* 2010. Paternally induced transgenerational environmental reprogramming of metabolic gene expression in mammals. *Cell* **143:** 1084–1096.

27. Bonnin, A. & P. Levitt. 2012. Placental source for 5-HT that tunes fetal brain development. *Neuropsychopharmacology* **37:** 299–300.

28. Bolcun-Filas, E. & J.C. Schimenti. 2012. Genetics of meiosis and recombination in mice. *Int. Rev. Cell. Mol. Biol.* **298:** 179–227.

29. Hartford, S.A. *et al.* 2011. Minichromosome maintenance helicase paralog MCM9 is dispensible for DNA replication but functions in germ-line stem cells and tumor suppression. *Proc. Natl. Acad. Sci. USA* **108:** 17702–17707.

30. Chang, G.Q. *et al.* 2008. Maternal high-fat diet and fetal programming: increased proliferation of hypothalamic peptide-producing neurons that increase risk for overeating and obesity. *J. Neurosci.* **28**(46)**:** 12107–12119.

31. Gross, G.A. *et al.* 1998. Opposing actions of prostaglandins and oxytocin determine the onset of murine labor. *Proc. Natl. Acad. Sci. USA* **95:** 11875–11879.

32. Plunkett, J. *et al.* 2010. Primate-specific evolution of noncoding element insertion into PLA2G4C and human preterm birth. *BMC Med. Genomics* **3:** 62.

33. Plunkett, J. *et al.* 2011. An evolutionary genomic approach to identify genes involved in human birth timing. *PLoS Genet* **7:** e1001365.

34. Hassan, S.S. *et al.* 2011. Vaginal progesterone reduces the rate of preterm birth in women with a sonographic short cervix: a multicenter, randomized, double-blind, placebo-controlled trial. *Ultrasound Obstet. Gynecol.* **38:** 18–31.

35. Romero, R. *et al.* 2012. Vaginal progesterone in women with an asymptomatic sonographic short cervix in the midtrimester decreases preterm delivery and neonatal morbidity: a systematic review and metaanalysis of individual patient data. *Am. J. Obstet. Gynecol.* **206:** 124.e1–e19.

36. Campbell, S. 2011. Universal cervical-length screening and vaginal progesterone prevents early preterm births, reduces neonatal morbidity and is cost saving: doing nothing is no longer an option. *Ultrasound Obstet. Gynecol.* **38:** 1–9.

37. Lely, A.T. *et al.* 2012. Circulating Lymphangiogenic Factors in Preeclampsia. *Hypertens Pregnancy* Sep 7. [Epub ahead of print].

38. Yin, Y. *et al.* 2012. Neonatal diethylstilbestrol exposure alters the metabolic profile of uterine epithelial cells. *Dis. Model Mech.* **19:** 870–880.

39. Ma, L. 2009. Endocrine disruptors in female reproductive tract development and carcinogenesis. *Trends Endocrinol. Metab.* **20:** 357–363.

40. Muhlhauser, A. *et al.* 2009. Bisphenol A effects on the growing mouse oocyte are influenced by diet. *Biol. Reprod.* **80:** 1066–1071.

41. Hunt, P.A. *et al.* 2012. Bisphenol A alters early oogenesis and follicle formation in the fetal ovary of the rhesus monkey. *Proc. Natl. Acad. Sci. USA* **109:** 17525–17530.

42. Wise, L.D. *et al.* 2009. Embryo-fetal developmental toxicity study design for pharmaceuticals. *Birth Defects Res. B Dev. Reprod. Toxicol.* **86:** 418–428.

43. Willett, C.E., P.L. Bishop, & K.M. Sullivan. 2011. Application of an integrated testing strategy to the U.S. EPA endocrine disruptor screening program. *Toxicol. Sci.* **123:** 15–25.

44. Birnbaum, L.S. *et al.* 2012. Consortium-Based Science: The NIEHS's Multipronged, Collaborative Approach to Assessing the Health Effects of Bisphenol A. *Environ Health Perspect* doi:10.1289/ehp.1205330

45. Fostel, J.M. 2008. Towards standards for data exchange and integration and their impact on a public database such as CEBS (Chemical Effects in Biological Systems). *Toxicol. Appl. Pharmacol.* **233:** 54–62.

46. Lim, H.J. & H. Wang. 2010. Uterine disorders and pregnancy complications: insights from mouse models. *J. Clin. Invest.* **120:** 1004–1015.